CHEMICAL ENGINEERING METHODS AND TECHNOLOGY

CAPILLARY ELECTROPHORESIS IN THE EARLY TWENTY-FIRST CENTURY

NEW TRENDS AND RELEVANT APPLICATIONS

CHEMICAL ENGINEERING METHODS AND TECHNOLOGY

Additional books and e-books in this series can be found on Nova's website under the Series tab.

CHEMICAL ENGINEERING METHODS AND TECHNOLOGY

CAPILLARY ELECTROPHORESIS IN THE EARLY TWENTY-FIRST CENTURY

NEW TRENDS AND RELEVANT APPLICATIONS

OSCAR NÚÑEZ
AND
XAVIER SUBIRATS
EDITORS

nova science publishers
New York

Library of Congress Cataloging-in-Publication Data

ISBN: 978-1-53615-223-4
Library of Congress Control Number: 2019934551

Published by Nova Science Publishers, Inc. † New York

CONTENTS

PREFACE

Capillary electrophoresis (CE) comprises a family of related separation techniques in which an electric field is used to achieve the separation of components in a mixture. One of the key features of CE is the simplicity of the instrumentation required, and today, these techniques have become powerful analytical tools for high efficiency separation, and the identification of a variety of ionic and neutral compounds.

Moreover, the versatility of CE operational modes (capillary zone electrophoresis, micellar electrokinetic capillary chromatography, capillary electrochromatography, capillary isoelectric focusing, isotachophoresis, non-aqueous capillary electrophoresis, etc.) and detection systems (ultraviolet, fluorescence, amperometric, mass spectrometry…) make CE useful for applications in many fields such as environmental, food, pharmaceutical, and forensic analysis. The typcial drawback of CE techniques, based on their low sensitivity when dealing with conventional UV-detection, due to their short detection pathways and the short sample volume introduced into the capillary, has been largely resolved not only by using more sensitive detectors like fluorescence, amperometric, or mass spectrometry, but by the employment of both off-line and on-line enrichment procedures.

The aim of this present book is to address the state-of-the art of capillary electrophoresis techniques within this early twenty-first century, by reviewing new trends and the most relevant applications described in the literature. Uses in fields such as food, environmental, forensic and biological analysis are addressed by means of relevant application. The scope of this book is intentionally broad and is aimed at worldwide analytical laboratories, both public and private, at acadamic institutions, as well as researchers.

Chapter 1 - Capillary electrophoresis involves a family of related separation techniques in which an electric field is used to achieve the separation of components in a mixture. Capillary zone electrophoresis (CZE) is one of the most widely used modes due to its simplicity of operation and versatility, and today is appearing as an alternative to other separation techniques, such as liquid chromatography, in multiple application fields. It is well known that CZE methods suffer from relatively high concentration detection limits due to the low sample injection volumes (few nanoliters in normal operation modes), and because the frequently employed UV detection is accomplished directly on the capillary with a short optical path length. Thus, for trace analysis applications, the amount of analyte introduced into the capillary or the detector sensitivity must be increased. The latter aspect may be accomplished by using light paths in connection with UV detection, or alternatively by using more sensitive detectors such as laser-induced fluorescence (LIF), or by means of coupling CE with mass spectrometry (CE-MS). Lately, one of the most frequently proposed approaches to improve CE detection is to increase the amount of analyte injected into the capillary by means of either by analyte-enrichment during sample preparation or by on-column electrophoretic-based preconcentration methods. All these approaches allow CE methods to easily reach trace level analysis and to be a reliable technique to fulfil the strong requirements of food safety and environmental analysis. In this book chapter, the role of capillary zone electrophoresis in food safety and environmental analysis will be discussed. Coverage of all kind of applications is beyond the scope of the present contribution, so the authors will focus on the most relevant applications published in the early twenty-first century by addressing the most interesting examples and important aspects in the food safety and environmental fields.

Chapter 2 - The characterization of food products based on compositional profiles of natural components obtained by capillary electrophoresis (CE) is discussed here. CE with different detection techniques, especially UV-visible spectroscopy and mass spectrometry, has widely been used for the generation of data to be further processed by chemometrics. It has been demonstrated that contents of food constituents are dependent on climatic, agricultural and manufacturing issues so they can be exploited as chemical descriptors of food features and quality. Chemometrics has widely been used as a way of facilitating the extraction of information dealing with food properties. In this regard, cluster analysis, principal component analysis and related methods are currently used for discrimination, classification, modelling and authentication purposes. The authors will comment on some representative examples to illustrate the possibilities of CE combined with chemometrics for the study of food products.

Chapter 3 - Forensic science is a strongly expanding research field and encompasses all the scientific disciplines that can be used to support law enforcement at any level. Just limiting the attention to the analytical aspect of forensic sciences, it involves an extremely wide application area that includes forensic toxicology, personal identification, analysis of explosives and gunshot residues, identification of fibers and hairs, inks and paints, etc. Consequently, the characterization and quantification of a wide range of compounds in a great variety of matrices must be addressed by forensic science. This fact supposes a big challenge faced by forensic scientists because the majority of forensic samples present a very complex nature and, moreover, their analysis requires very selective and sensitive analytical techniques. In the last decades, capillary electrophersis has shown to be a powerful tool that provides solutions to almost any analytical question arising in forensic laboratories. As result, it has become an complementary and essential separation technique in this field, particularly for the analysis of forensic DNA. To illustrate this fact, this chapter offers a comprehensive overview of the most recent applications of CE in forensic analytical cemistry, covering its different separation modes and its hyphenation with different detection systems.

Chapter 4 - The vast majority of drugs are chiral, thus, they can exist as two enantiomers. In most cases enantiomers display different pharmacological properties. Given the increasing demand on commercializing drugs in the enantiomerically pure format, there is the requirement to develop methodologies capable of resolving enantiomers. In this aspect, capillary electrophoresis has attracted much interest, thanks to its numerous advantages such as high efficiency, minimal sample and solvent consumption as well as its elevated flexibility in modifying the chiral environment, which increases the chances of achieving a chiral separation. The objective of this Chapter is to provide an insight on the recent contributions of capillary electrophoresis to the chiral analysis of drugs published from January 2016 to March 2018. All reported contributions used UV detection being, in almost all cases, the electrokinetic chromatography mode using cyclodextrins, the strategy of choice.

Chapter 5 - Capillary electrochromatography (CEC) is an analytical technique performed in capillaries with a thin internal diameter containing a stationary phase (SP). Compounds are moved into the column to the detector by a strong electroosmotic flow (EOF) generated by the application of a high voltage (10-20 kV). The separation mechanism involved in the analysis process is based on both chromatographic and electrophoretic principle. Therefore it offers high efficiency and high selectivity typical of capillary electrophoresis and high performance liquid chromatography (HPLC), respectively. CEC has been widely studied and the literature covers several aspects related to this miniaturized technique such as theory, instrumentation, selection and use of different stationary phases, applications etc. The aim of this chapter is to present the main features of CEC and its potentiality for the separation of compounds of different nature (uncharged and charged species). Instrumentation currently used considering the detectors and the coupling with mass spectrometry (MS), optimization of experimental conditions, the different SPs have also been discussed. Finally some selected applications in the field of drug analysis, protein, enantiomers, nutraceutical compounds separation are described.

Chapter 6 - In this chapter chromatographic methods are examined for lipophilicity determination and as surrogate models for biological activity, particularly micellar and microemulsion electrokinetic chromatography (MEKC and MEEKC). The fundamentals of these techniques are presented, together with the relationships of lipophilicity or biological activity with the mass distribution ratios (or capacity factors) obtained from the migration of analytes in the capillary electrophoretic system. The Abraham's solvation model, based on linear free energy relationships, is presented as a very powerful tool for the characterization of partition processes and their comparison in order to find the best possible matches between chromatographic, physicochemical and biological systems.

Chapter 7 - Capillary Electrophoresis (CE) has led a turbulent life in the pharmaceutical industry. Application of the CE techniques have continuously been increasing, but after the initial enthusiastic phase and fast adaptation, there was a decline in use at the start of the 21st century for small molecules. In this chapter the implementation and use of the CE applications in the pharmaceutical and biotechnology industry in the early twenty-first century are discussed. The requirements for CE to become a well-established technique for the analysis of therapeutic proteins are discussed as they also set the requirements for further development. Relevant old and new applications in industry illustrate the versatility of use of the CE techniques in industry. The application of Analytical Quality by Design (AQbD) on capillary electrophoresis method development, validation, implementation, and life cycle management is reviewed. The chapter ends with expectations of and hopes for the future use of the CE techniques. If the CE community keeps sharing good practice and good applications, CE will have a bright future in the pharmaceutical and biotech industry.

In: Capillary Electrophoresis … ISBN: 978-1-53615-223-4
Editors: O. Núñez and X. Subirats

Chapter 1

THE ROLE OF CAPILLARY ZONE ELECTROPHORESIS IN FOOD AND ENVIRONMENTAL ANALYSIS

Oscar Núñez[1,2,3,*] and Xavier Subirats[1]

[1]Department of Chemical Engineering and Analytical Chemistry, University of Barcelona, Barcelona, Spain
[2]Research Institute in Food Nutrition and Food Safety, University of Barcelona, Barcelona, Spain
[3]Serra Húnter Fellow, Generalitat de Catalunya, Spain

ABSTRACT

Capillary electrophoresis involves a family of related separation techniques in which an electric field is used to achieve the separation of components in a mixture. Capillary zone electrophoresis (CZE) is one of the most widely used modes due to its simplicity of operation and versatility, and today is appearing as an alternative to other separation techniques, such as liquid chromatography, in multiple application fields. It is well known that CZE methods suffer from relatively high

* Corresponding Author's Email: oscar.nunez@ub.edu.

concentration detection limits due to the low sample injection volumes (few nanoliters in normal operation modes), and because the frequently employed UV detection is accomplished directly on the capillary with a short optical path length. Thus, for trace analysis applications, the amount of analyte introduced into the capillary or the detector sensitivity must be increased. The latter aspect may be accomplished by using light paths in connection with UV detection, or alternatively by using more sensitive detectors such as laser-induced fluorescence (LIF), or by means of coupling CE with mass spectrometry (CE-MS). Lately, one of the most frequently proposed approaches to improve CE detection is to increase the amount of analyte injected into the capillary by means of either by analyte-enrichment during sample preparation or by on-column electrophoretic-based preconcentration methods. All these approaches allow CE methods to easily reach trace level analysis and to be a reliable technique to fulfil the strong requirements of food safety and environmental analysis.

In this book chapter, the role of capillary zone electrophoresis in food safety and environmental analysis will be discussed. Coverage of all kind of applications is beyond the scope of the present contribution, so we will focus on the most relevant applications published in the early twenty-first century by addressing the most interesting examples and important aspects in the food safety and environmental fields.

Keywords: capillary zone electrophoresis, food analysis, environmental analysis

INTRODUCTION

Capillary electroforesis (CE) emerged as a very exciting and promising analytical separation technique and CE instruments became commercially available by the late 1980s–mid 1990s. CE is a well-recognized separation technique that is based on the differential migration of charged chemical species within narrow diameter silica capillaries (20 to 200 μm) filled with a background electrolyte under the influence of a high electric field strength generated by applying a high voltage between the capillary tips. One of the key features of CE is the simplicity of the instrumentation required, and today this technique allows working in various modes of operation, which includes capillary zone electrophoresis (CZE), micellar electrokinetic capillary chromatography (MECC), capillary gel electrophoresis (CGE), capillary isoelectric focusing (CIEF), capillary isotachophoresis (CITP), non-aqueous capillary electrophoresis (NACE), and capillary electrochromatography (CEC), among others.

Capillary zone electrophoresis is the most widely used CE mode due to its simplicity of operation and its versatility. The employment of high electric field strengths results in short analysis time and high efficiency and resolution. In addition, the minimal sample volume required (generally, few nanoliters), the on-capillary detection, the potential for both qualitative and quantitative analysis, the automation, and the possibility of hyphenation with other techniques such as mass spectrometry (MS) has allow CZE to become one of the premier separation techniques in multiple fields such as bio-analysis [1–3], food control and safety [4–7], and even in environmental applications [8] where low analyte concentration levels are expected.

Nowadays, the benefits from the high number of theoretical plates obtained with CZE have been overshadowed by the poor sensitivity and, consequently, the low limits of detection achieved with this technique when UV-visible detection is employed, and today this is one of the main handicaps of CE techniques. Due to the small dimensions of CZE capillaries, tipically in the 20-200 μm internal diameter (I.D.) range and the capillary lengths employed in most of the applications (40-80 cm total length), the sample volumes that are loaded into the capillary are very small (few nL). For example, in a 50 μm I.D. x 50 cm total length capillary only 1.18 nL/s of sample are injected into the capillary when hydrodynamic injection (0.5 psi) is employed. Moreover, when the most common optical detection techniques are used, CZE suffers from a drastically reduced path length in comparison to liquid chromatography (LC) techniques due to the fact that detection is carried out directly in the capillary, being its internal diameter the optical path length. As UV-visible absorbance is directly proporcinal to the used path length and the analyte concentration in the sample, the concentration of the samples must be dramatically increased to obtain the same signal-to-noise ratio as would result from a typical LC analysis.

Additionally, environmental and food safety analysis frequently requires the determination of organic contaminants, which can be numbered in the thousands, and that are present at extremely low concentrations in very complex matrices. Any instrumental technique that is to be used successfully in environmental and food analysis should achieved the separation of the matrix compounds and the identification of the contaminants at trace analysis [9]. Thus, when trace analysis applications are required, the amount of analyte to be introduced into the CE capillary must be incresed or the sensitivity of the detection system employed must be improved. The latter aspect can be accomplished by using light paths in connection with UV detection, in order to increase the path length during UV-detection [10], or by using more sensitive detectors such as laser-induced fluorescence (LIF) detection [8, 11–13], amperometric detection [14, 15] or the coupling of CZE with mass spectrometry (CE-MS) techniques [16–20]. Both bubble cells and z-shaped cells have been employed as extended path lengths in UV detection, which typically provide 3 to 5-fold signal-to-noise enhancements, although this is not enough for some specific trace-analysis applications. High mass sensitivity and selectivity have been described when LIF detection is employed, but the main handicap of this detection system relies on its selectivity, because it is only applicable for those compounds with fluorescence properties, and the number of wavelengths available with the commercial available LIF detectors is still limited. Regarding the coupling with mass spectrometry, CE-MS techniques are becoming promising techniques and are frequently employed to increase sensitivity, but this coupling requires specialized interfaces, and frequently this technique is combined with preconcentration methods.

Obviously, sample preconcentration procedures can be employed to improve sensitivity in any analytical technique. The most common sample treatment and preconcentration strategies typically used in environmental and food safety analysis involved liquid-liquid extraction (LEE) [21] and solid-phase extraction (SPE) [22]. However, although good sensitive enhancements can be achieved with these methods, they are time consuming (espacially when addressed in an off-line configuration), and will not be discussed in this chapter.

The most convenient approach to improve sensitivity in CZE is by increasing the amount of analyte to be introduced into the capillary by on-column electrophoretic-based preconcentration methods [23, 24]. The use of this approach does not require any special instrument set-up configuration, which is the reason why is one of the most frequently proposed strategies to improve sensitivity in CZE when UV-visible detection is used. These methods mainly involve manipulating the electrophoretic velocity of the analyte during injection and separation on CZE, and include techniques such as normal sample stacking, large-volume sample stacking (LVSS), field-amplified sample injection (FASI) and electrokinetic supercharging (EKS), among others.

In the next sections, the role of capillary zone electrophoresis in environmental and food safety analysis will be addressed. We do not intent to cover all kind of applications so the present contribution will focus on some relevant application examples published within the earlier twenty-first century.

FOOD ANALYSIS

Food products are very complex mixtures constituted not only of naturally occurring compounds but of many other chemicals (some of them contaminants) coming from agrochemical treatments, technological processes or even migrating from packaging materials. Food safety is ensured through the legislative framework that relies on appropriate analytical methods to detect contamination of food, both rapidly and reliable. The analysis of organic contaminants in food products is an important issue that involve many disciplines including chemistry, biochemistry and microbiology. Moreover, it is a difficult task because of the huge number of contaminants (numbered in the thousands) coming from different sources that can be present in the final foodstuff, but also due to the fact that many of them are present at very low concentration levels in a matrix with natural major constituents that can produce an important matrix effect during the analysis. The analytical techniques addressing the analysis of food products should then achieve the separation of matrix components that can interfere during the analysis, and the identification and quantitation of the targeted contaminants even at trace levels.

Nowadays, CZE has emerged as a powerful separation analytical technique in the routine food analysis, and several review articles devoted to this topic can be found in the literature [4, 7, 9, 25–27]. Targeted multiresidue methods, screening strategies in food routine analysis, food safety, food integrity, food traceability, and metabolomics approaches in food authentication (Foodomics) are among the application fields where CZE may be considered a very well suited technique, especially for analytes that are neither determinable by gas chromatography (GC) or cannot be separated by liquid chromatography (LC). Some examples centered on the topics discussed above will be presented below.

Regarding food safety issues, one field where CZE has demonstrated to be one of the techniques of choice during the early twenty-first century is in the analysis of veterinary drugs [15, 17, 18, 20, 28–38]. On modern farms, animals are raised under confined-feeding conditions at high temperature and humidity that can easily lead to the proliferation of microorganisms causing many diseases. As a consequence, the prevention and treatment of these illnesses via the administration of a varied range of veterinary drugs is required. However, residues of the drugs, or even some metabolites, can remain in the animal tissues or in food-derived animal products such as milk, eggs, etc.; hence, the drug or their metabolites (sometimes even more toxic than the native drug) can arrive to the consumers if withdrawal periods are not fully accomplished.

Several multi-residue methods based on CZE for the determination of veterinary drugs in food are described in the literature. Due to the low concentration levels at which these drugs need to be monitored to comply legislation most of these methodologies requires the use of enrichment procedures in order to improve sensitivity. On-column electrophoretic-based preconcentration methods [23, 24, 39] emerge as the most appropriate in these cases because they do not need any special instrument set-up configuration, and only requires the manipulation of the electrophoretic velocity of the targeted compounds during injection and the separation steps. In this sense, large-volume sample stacking [31] and field-amplified sample stacking (FASS) [33, 34] are typically employed in the multi-residue analysis of veterinary drugs in foodstuffs by capillary electrophoresis. For example, Li et al. [33] proposed the multi-residue analysis of sulfonamides in chicken meat by means of a poly(methacrylic acid-*co*-ethylene glycol dimethacrylate) monolith microextraction (PMME) and on-line preconcentration using FASS-CZE. The PMME allowed to improve extraction efficiency of sulfonamides and to reduce matrix interferences from chicken samples. By using FASS, limits of detection of 3.49-16.7 ng/g were achieved, with very satisfactory precision values (RSD below 13%) and recoveries (96.3-104%) over a linear range of 50-1000 ng/g. A similar approach was employed by the same research group in the multi-residue determination of (fluoro)quinolone antibiotics also in chicken samples [34].

Another approach to address multi-residue CZE methods is by employing more selective and sensitive detection systems such as mass spectrometry [17, 18]. For instance, Lara et al. [17] proposed a capillary zone electrophoresis-tandem mass spectrometry (CZE-MS/MS) method for the identification and simultaneous determination of eight quinolones for veterinary use in bovine raw milk. MS/MS experiments using an ion trap as analyzer operating in the multiple reaction monitoring mode were carried out to achieve the minimum number of identification points according to the 2002/657/EC European Decision [40].

Electrospray (ESI) was employed as ionization source by means of a coaxial sheath-liquid sprayer. For that purpose, a sheath liquid consisting of 2-propanol/water/formic acid (50:49:1 v/v/v) was employed. For the quantitation of these compounds in such a complex matrix like raw milk, the authors employed a two-step SPE procedure using Oasis MAX and HLB cartridges without protein precipitation. As an example, Figure 1 shows the CE-MS/MS electropherograms obtained under optimum conditions after the two-step SPE procedure of milk samples spiked with quinolone antibiotics at different concentration levels. The proposed method showed very satisfactory results in terms of linearity (determination coefficient r^2 higher than 0.989) and precision with relative standard deviation (RSD) values below 16%.

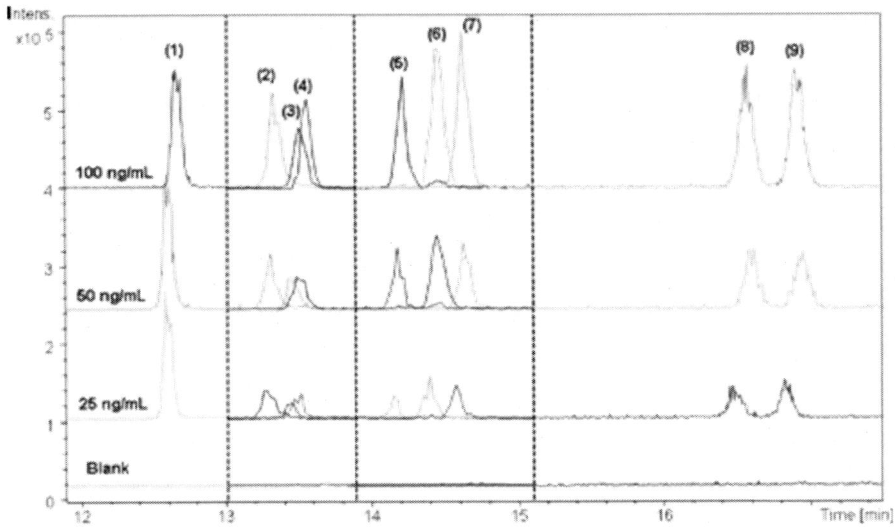

Figure 1. CE-MS/MS electropherograms at optimum conditions obtained after SPE of milk samples spiked with quinolone antibiotics at different concentration levels. Peak identifications: (1) Lomefloxacin (1 μg/L, used as internal standard), (2) Danofloxacin, (3) Sarafloxacin, (4) Ciprofloxacin, (5) Marbofloxacin, (6) Enrofloxacin, (7) Difloxacin, (8) Oxolinic acid, and (9) Flumequine. Reprinted with permission of reference [17]. Copyright (2006) American Chemical Society.

The limits of detection (LOD) and quantitation (LOQ) values (below 6 and 24 μg/L, respectively) achieved with the proposed method were in all cases lower than the maximum residue limits (MRL) tolerated for these family of compounds in milk products, and the observed recoveries ranged from 81 to 110%, showing the potential of CZE-MS/MS for the analysis of regulated quinolone antibiotics in food quality and safety control laboratories.

The complexity of food matrices can produce interferences in the target analysis of organic contaminants in foodstuffs. In those cases, the use of mass spectrometry techniques is required to identify the interfering signal. For example, Martínez-Villalba et al. [35] proposed and validated a CZE-UV method to analyze residues of amprolium, a quaternary ammonium compound used in the treatment of coccidiosis, in eggs. In order to improve sensitivity and reach the MRL legislated, field-amplified sample injection (FASI) was used for in-line preconcentration after a quick and simple sample treatment based on solid-phase extraction (SPE) using Envi-Carb cartridges. However, during method-validation studies using egg-samples, the authors encountered the presence of a matrix interference at the migration time of amprolium, preventing the correct quantification of this veterinary drug.

In order to identify the interfering compound the authors employed CZE in combination with mass spectrometry with an ion-trap mass analyzer, being able to confirm by MS^n experiments the identity of the interfering signal as thiamine (Vitamine B1), as can be seen in Figure 2. The authors reoptimized the CZE separation conditions to achieve the baseline separation of thiamine from amprolium (Figure 3) allowing the quantification of amprolium in eggs at concentrations down to 75 μg/L, which were far below the MRL-legislated values.

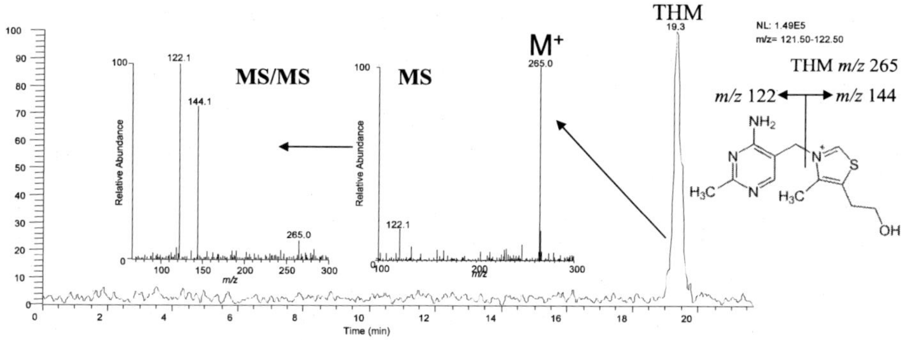

Figure 2. Electropherogram of thiamine and its MS and MS/MS spectra. Reprinted with permission of reference [35]. Copyright (2013) Wiley-VCH.

Another interesting example where CZE played an important role in the food safety area is in the analysis of melamine. In September 2008, infant formulas that were illegally adulterated with melamine (MEL) led to health problems for thousands of infants in China. This tragic episode captured attention from all over the world. MEL is typically used to make plastics, fertilizers, and other products but its use as a food ingredient has never been authorized. Nevertheless, nitrogen-rich MEL was deliberately added to food products because protein content is usually estimated by determining the nitrogen content, and the added MEL can boost the nitrogen content of the foodstuff so as to make them appear to have more protein and to reduce the production costs.

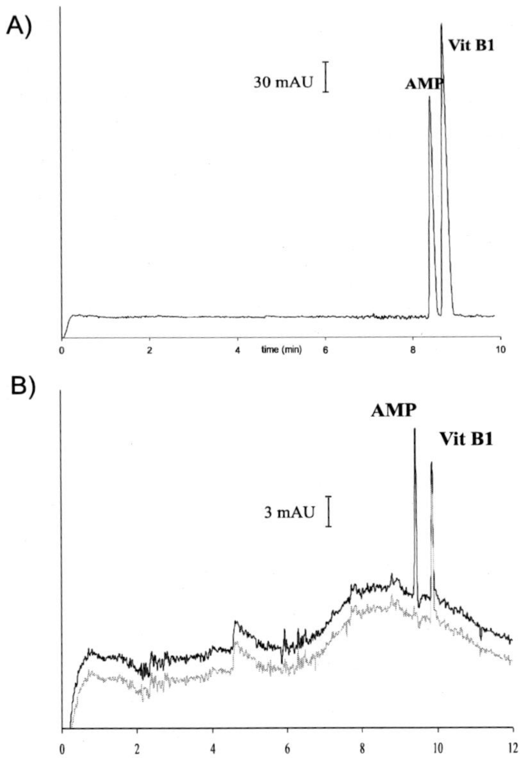

Figure 3. (A) Separation of Amprolium and Thiamine by CZE-UV with 150 mM acetic acid-ammonium acetate pH 5.0:40% MeOH buffer at 30 kV separation voltage. (B) Blank egg sample spiked at 100 µg/kg overlayed to a blank egg sample. Reprinted with permission of reference [35]. Copyright (2013) Wiley-VCH.

Therefore, at that moment there was an urgent need to establish simple and effective methods for the determination of MEL in food products, and several CZE methods can be found in the literature [41–43].One of the first papers dealing with the analysis of melamine by CZE in foodstuffs was published by Yan et al. [41]. The authors described the determination of this compound in liquid milk, yogurt, whole milk powder, fish feed, and fish at residue levels by employing CZE with diode array detection an using a 30 mM NaH_2PO_4 solution at pH 3.2 as running buffer. Samples suspicious to be adulterated with MEL were extracted with 1% trichloroacetic acid (TCA) and 1 mL chloroform was employed to remove fat by precipitation. Figure 4 shows the electropherogram of a standard of MEL in 1% TCA (a), and the electropherograms of the analyzed samples, as well as those corresponding to the sample spiked with 2μg/L MEL: (b) liquid milk, (c) yogurt, (d) whole milk powder, (e) fish powder, and (f) fish. As can be seen, CZE-DAD running time was below 6 min. Total analysis time including both sample preparation and CZE-DAD analysis could be completed within 30 min. The limits of detection and quantitation for MEL were 0.01 and 0.05 μg/mL, respectively. The authors successfully applied the proposed method for the analysis of MEL in dairy products, fish feed and fish with total recoveries ranging from 93 to 104%.

Regarding food safety issues, other application fields where CZE is playing an important role is in the analysis of pesticides, and several examples can be found in the literature [44–48]. Obviously, in this kind of applications and in order to address the low sensitivity typically observed with conventional UV-detection, preconcentration strategies are required. For example, Aulakh et al. [45] proposed the use of SPE using C18 and polymeric sorbent cartridges for the preconcentration of quaternary ammonium herbicides in beverages by CZE, while Quesada-Molina et al. [48] employed an on-line electrophoretic-based preconcentration method such as LVSS for the trace determination of sulfonylurea herbicides in water and grape samples by CZE. In other cases, the use of more selective and with higher sensitivity detectors are proposed, such as for instance electrochemical detection [46, 47]. In that sense, Lin et al. [46] addressed the determination of phenoxyl-type N-methylcarbamate pesticide residues in vegetables by CZE with pre-column hydrolysis and amperometric detection. This method showed to be 20 to 50-fold more sensitive than those previously reported with conventional UV-detection.

The determination of mycotoxins, secondary metabolites produced by fungal species growing on agricultural products during cultivation, harvest, transport and storage [49], is also an important food safety issue, and several CZE methodologies can be found in the literature addressing these analyses [50, 51]. For example, a simple analytical methodology based on CZE for the screening and quantification of zearalenone and its metabolites, α-zearalenol and β-zearalenol, was described in the literature [51].

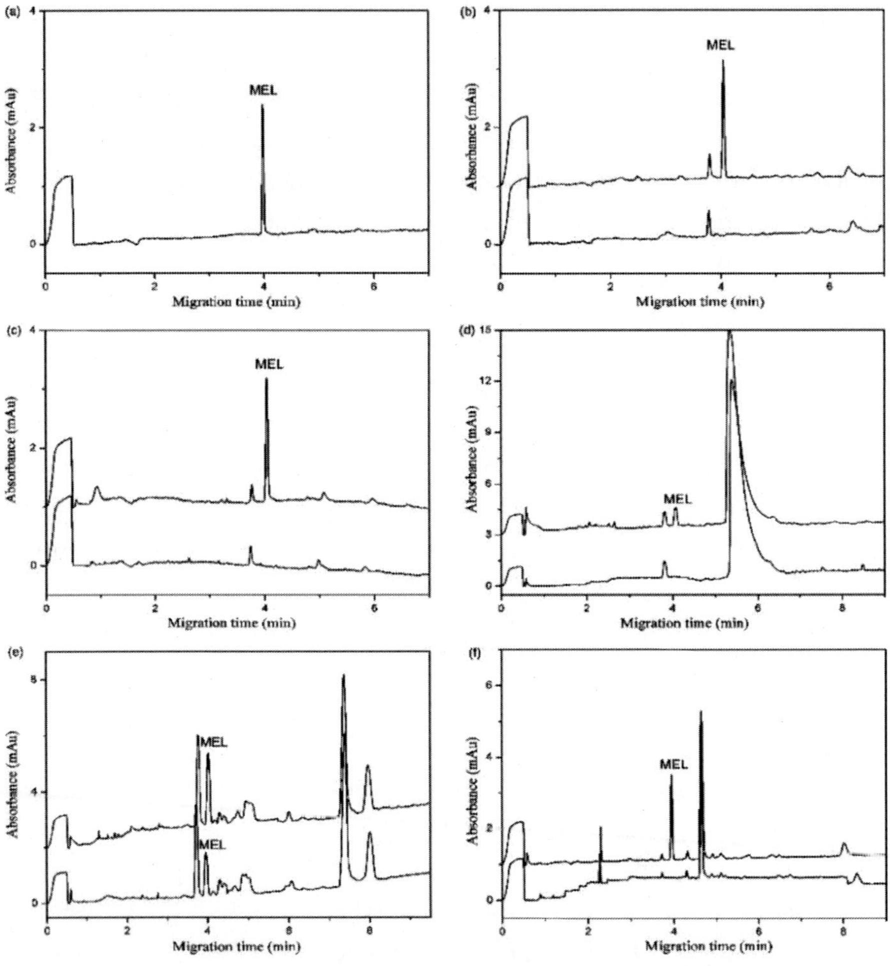

Figure 4. Typical electropherograms of an standard, samples, and spiked samples under optimal conditions: (a) standard of MEL in 1% TCA; (b) liquid milk and spiked liquid milk; (c) yogurt and spiked yogurt; (d) whole milk powder and spiked whole milk powder; (e) fish feed and spiked fish feed; (f) fish and spiked fish. The upper lines in b-f figures correspond to the real samples spiked with 2 μg/mL MEL. Reprinted with permission of reference [41]. Copyright (2009) American Chemical Society.

Maize flour extracts obtained by supercritical fluid extraction were analyzed by CZE (using 25 mM borate separation buffer at pH 9.2 and 25 kV as separation voltage) with amperometric detection (monitoring the signal at +700 mV with a carbon paste electrode). This scheme allowed a rapid and reliable identification of contaminated flour samples according to the reference value established by legislation (200 μg/kg). The method was able to process samples in 4 min with detection limits of 12 μg/L, enough to discriminate between positive (more than 200 μg/L total mycotoxins) and negative samples.

CZE was also employed in the determination of heat-induced contaminants such as acrylamide and heterocyclic amines in food products [28, 29, 52, 53], or in the determination of furosine, a well-known marker of the extent of thermal treatment applied to foodstuffs [54]. As an example, Bermudo et al. [28, 29] proposed the use of CZE for the determination of acrylamide after derivatization with 2-mercaptobenzoic acid in different foodstuff products such as homemade French fries, breakfast cereals and biscuits. FASI and LVSS preconcentration methods were also described in these applications to enhance CZE sensitivity [29].

Food integrity and authenticity is an emerging food topic in our society where capillary zone electrophoresis is playing also an important role in the last years. Although food adulteration has a long story, recent scare events such as the melamine adulteration previously commented have focused media attention on this area. The deliberate adulteration of food and misrepresentation to deceive the final consumer is illegal world-wide. As the food supply chain is nowadays a global issue, is exceedingly complex and there are many players involved between production and final consumption, it is in fact much easier to conduct fraud without being easily detected. In addition, today society is increasingly interested in the consumption of food products with certain quality attributes such as the presence of bioactive substances (functional foods), with certain organoleptic characteristics, produced following certain agricultural practices (organic products), or grown in specific regions (products with protected designation of origin, PDO). Authentication of these products is also necessary because fraudulent practices will not have only economic consequences but can also represent important health issues when prohibited additives are added to deceive the organoleptic properties of the final adulterated food product, or when the adulterant can produce acute allergy episodes. The application of capillary zone electrophoresis in the field of food characterization, classification and authentication is gaining interest [55–65].

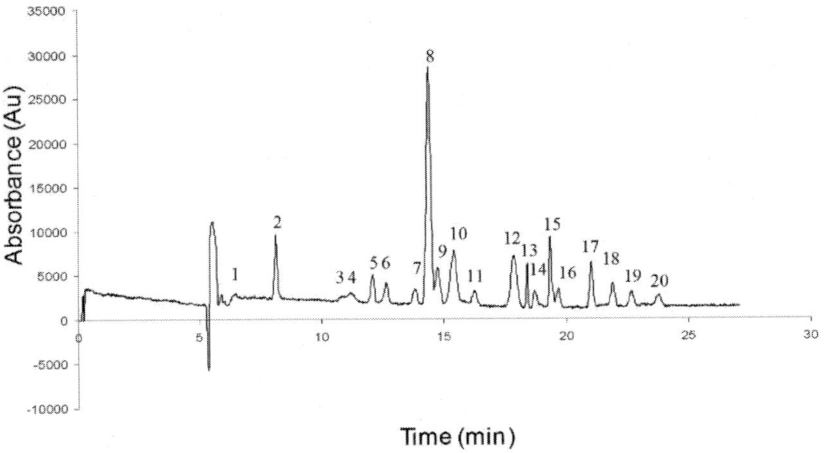

Figure 5. Electrophoretic separation of an aqueous standard mixture of 20 polyphenols. Running buffer, 30 mM tetraborate buffer with 5% isopropanol; capillary voltage, +25 kV; pressure-assisted separation (3.5 kPa) from minute 18; acquisition wavelength, 280 nm. Peak identification: (1) 1-(4-hydroxyphenyl)ethanol, (2) resveratrol, (3) (-)-epicatechin, (4) (+)-catechin, (5) veratric acid, (6) homovanillic acid, (7) vanillin, (8) *trans*-cinnamic acid, (9) sinapic acid, (10) quercitrin, (11) homogentistic acid, (12) syringic acid, (13) ferulic acid, (14) fisetin, (15) *p*-coumaric acid, (16) quercetin, (17) 4-hydroxybenzoic acid, (18) caffeic acid, (19) gallic acid, and (20) 3,4-dihydroxybenzoic acid. Reprinted with permission of reference [60]. Copyright (2012) American Chemical Society.

Vallejo-Cordoba et al. [59] presented an overview of different meat authenticity issues solved by means of unique CE applications. CE methods based on meat protein fingerprinting are discussed by the authors for the analysis of meat species in unheated meat products. For protein data interpretation, patter recognition is used to account for the natural variability present within the same meat species. Protein-based CE methods or by monitoring specific markers are employed for the determination of meat extenders. For example, polyphenols are used as specific markers for soy detection and hydroxyproline is used as a specific marker for collagen determination.

Franquet-Griell et al. [60] developed a CZE method for the simultaneous determination of 20 polyphenols in wine samples. The separation was achieved using fused-silica capillaries of 75 μm i.d. and a 30 mM sodium tetraborate buffer solution at pH 9.2 with 5% isopropanol a running buffer. By employing a capillary voltage of +25 kV with pressure-assisted (3.5 kPa) separation from minute 18, a total analysis time below 25 min was obtained (Figure 5). LOD values between 0.3 and 2.6 mg/L, with good linearity and run-to-run and day-to-day precision values (RSD below 15.7%) were achieved. The proposed method was then applied for determination of polyphenols in 49 Spanish wines, showing very satisfactory results and a wide compositional variation between wines.

Taking into account that in natural food products the content and distribution of bioactive substances such as polyphenols may be attributed to many parameters such as geographical areas of production, water resources, environment conditions, growing practices, etc., the obtained electrophoretic profiles and other composition data (e.g., peak areas of selected peaks) was considered by the authors as fingerprints of wines to be used for characterization and classification purposes. Fifty-three Spanish wines from three different regions (Catalunya, La Rioja, and Castilla-La Mancha) were analyzed. As an example, Figure 6a shows the electropherograms of a wine sample recorded at different acquisition wavelengths. The fifteen peak signals selected by the authors for a non-supervised principal component analysis (PCA) are indicated, and the PCA results (score and loading plots) are given in Figure 6b. PCA results showed that the first principal component (PC1) was mainly focused on the description of the peak intensities, and variance dealing with geographical characteristics was not retained.

Information of the origin of wines was captured by PC2 and PC3, and as can be seen in Figure 6b, good discrimination amount the analyzed wine samples regarding the production region was achieved, with wines from Catalunya located on the right part of the score plot, wines from La Rioja on the top and central-left part of the plot, and wines from Castilla-La Mancha located mainly on the left side.

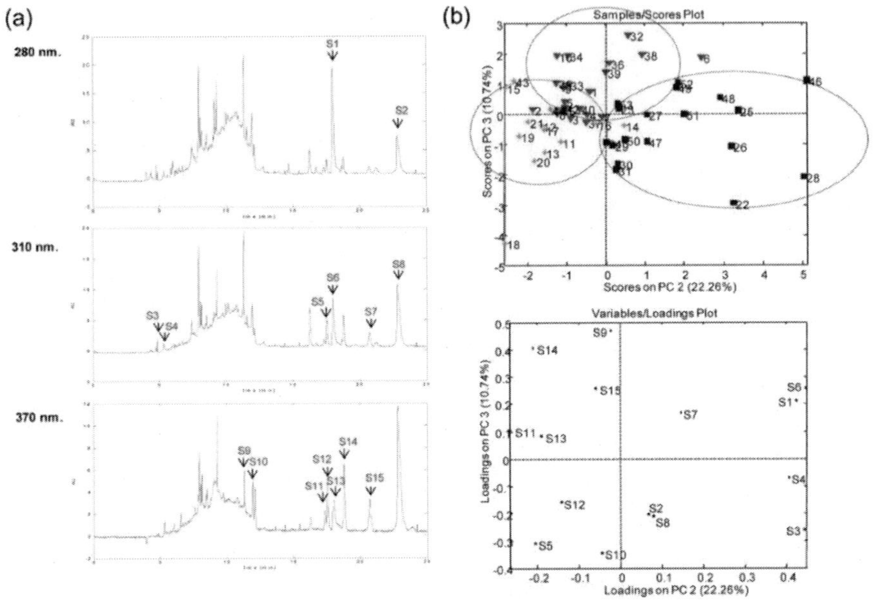

Figure 6. (a) Electropherograms of a wine sample recorded at 280, 310 and 370 nm. (b) PCA results (score and loading plots) using selected peak areas as analytical data. Reprinted with permission of reference [60]. Copyright (2012) American Chemical Society.

CZE profiling of bioactive substances in food products has also been described as a good strategy to address food integrity and authenticity issues. For example, Navarro-Pascual-Ahuir et al. [61] developed a method for the rapid differentiation of commercial juices and blends employing sugar-profiling by CZE with indirect UV-detection, which was achieve by using 10 mM 2,6-pyridine dicarboxylic acid (PDC) as indirect anionic chromophore in a 0.5 mM cetyltrimetrylammonium bromide at pH 12.1 running buffer. The authors evaluated different chemometric methods based on sugar content ratios. The best results were obtained with linear discriminant analysis (LDA), which showed that fruit juices can be distinguished according to the fruit type, being able even to differentiate juice blends, as can be observed in Figure 7.

In order to evaluate the possibility of detecting different percentages of fruit juices in binary mixtures by employing sugar profiling obtained by CZE, the authors constructed two multiple linear regression (MLR) models for pineapple-grape and orange-grape mixtures. Excellent prediction capabilities on the amount of adulterant (grape juice) were obtained, with average validation errors below 5.2%, showing that the proposed methodology provided an effective tool for quality control purposes in the juice industry and regulatory agencies.

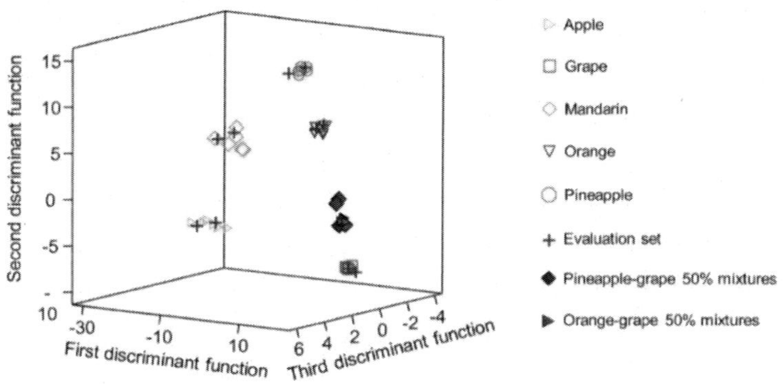

Figure 7. Score plot on an oblique plane of the three-dimensional space defined by the three first discriminants functions of the LDA model constructed to discriminate between juices obtained from different fruits. Evaluation set samples and fruit juice blends at a 50% level of each juice are labeled as indicated in the figure legend. Reprinted with permission of reference [61]. Copyright (2015) American Chemical Society.

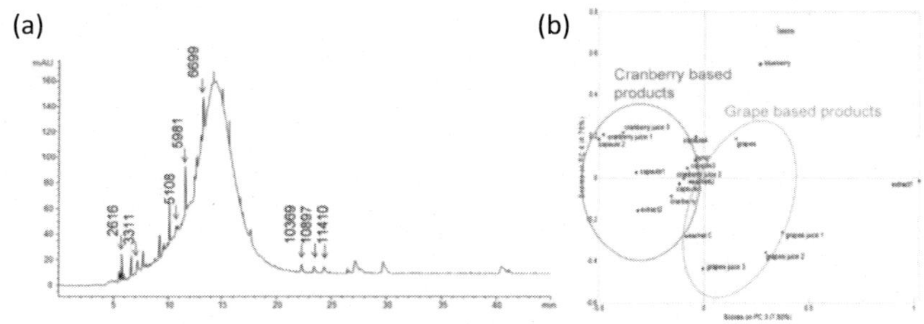

Figure 8. (a) Electropherogram obtained for a cranberry commercial capsule. Discriminant peak signals are indicated with an arrow. (b) PCA result (PC3 vs PC4 score plot) using eight discriminant peak signal areas. Adapted with permission of reference [65]. Copyright (2014) American Chemical Society.

In another work, polyphenolic profiling based on CZE-UV fingerprinting of fruit products was employed for the authentication of cranberry-based natural extracts [65]. CZE separation was performed by using a 35 mM sodium tetraborate aqueous solution (pH 9.4) containing 5% (v/v) methanol. The obtained CZE-UV polyphenolic fingerprints, as well as some selected discriminant signals, were employed to perform principal component analysis. As an example, Figure 8 shows the discriminant peak signals employed in a CZE-UV fingerprint of a cranberry commercial capsule sample (a) and the corresponding PCA score plot.

As can be seen, grape-based products were located on the right part of the plot, perfectly discriminated from cranberry-based products which appeared on the left side, grouped independently of the kind of product (fruit, juice, syrup, natural extract, capsule, or sachets). Sample products belonging to other fruits, such as blueberry and raisins, were plotted far away from the two main groups corresponding to grape- or cranberry-based products. An interesting result of this work is that a cranberry-based pharmaceutical extract (extract 1) is not located in the area corresponding to cranberry-based products, being likely to be a fraudulent sample. This hypothesis was confirmed by the authors by performing a high resolution mass spectrometry (HRMS) analysis of the extract using an Orbitrap MS instrument. The HRMS analysis reported a lack of presence of A-type proanthocyanidins compounds in extract 1 allowing the authors to confirm that this extract was not mainly a cranberry-based extract, as suspected. Thus, the use of CZE-UV fingerprinting profiling combined with PCA (selecting the most relevant peak signals in CZE) resulted in a cheap, straightforward, and useful strategy for the authentication of natural fruit-based extracts according to the fruit of origin.

ENVIRONMENTAL ANALYSIS

As previously commented, CZE is one of the most widely used mode in CE due to its simplicity of operation and its versatility, and today is appearing as an alternative to other separation techniques such as liquid chromatography in multiple application fields. Despite the low sensitivity characteristic of CZE when conventional UV-detection methods are employed, this technique has been proposed for multiple environmental applications. The first environmental applications of capillary zone electrophoresis were focused on the analysis of inorganic anions [66, 67]. However, in recent years capillary zone electrophoresis has begun to play an important role in the environmental determination of numerous organic pollutants thanks to the use of pre-concentration methods (both off-line and on-line approaches [68–70]) and the use of more selective and sensitive detectors such as electrochemical detection, fluorescence detection or mass spectrometry [8]. Several application examples will be presented below.

Laser-induced fluorescence is among the most selective and sensitive detection system employed in combination with CZE for the analysis of organic contaminants in environmental samples. For example, Chen et al. [11] employed multiphoton-excited fluorescence by diode laser of continuous wave for CZE to determine aniline species metabolized from pesticides. In this application, in order to achieve 2-photon excitation fluorescence, a derivatization procedure was performed using fluorescein isothiocyanate (FITC). Under optimal conditions, four pesticide residues were separated and determined within 4 min, and achieving a detection limit down to zepptomole level (calculated detection volume of 45.0 aL). Quantitative analysis also exhibited excellent linear dynamic relationship in a range of two orders of magnitude. Recently, the sensitive determination of sulfonamides in environmental waters by CZE coupled with both silvering detection window and in-capillary optical fiber light-emitting diode-induced fluorescence detection was proposed [12]. The strategy of this new detector was that half the surface of the detection window was coated with a silver mirror, which could reflect the undetected fluorescence to the photomultiplier tube to be detected, consequently enhancing the detection sensitivity. With this improved detection system, several sulfonamides in environmental water samples were determined achieving LODs in the range of 1.0-2.0 nM, with good intra-day and inter-day precisions.

In a similar way as in the case of the analysis of food samples, capillary zone electrophoresis coupled to mass spectrometry is gaining also popularity in the analysis of environmental samples and the number of applications that can be found in the literature is increasing every year [71]. However, although higher sensitivity is frequently obtained with CE-MS in comparison to conventional CZE-UV methodologies, preconcentration methods are frequently required to achieve the required legislated values for the determination of common organic pollutants. For example, Safarpour et al. [72] developed a method for the quantitative analysis of imazamox herbicide in environmental water samples by CE-MS with electrospray ionization. After a SPE sample treatment processing 1L of sample, limits of detection and quantitation of 200 and 20 ng/L, respectively, were obtained. In a very recent work, Zhang et al. [73] combined an off-line preconcentration procedure based on immunoaffinity extraction with CE-MS for the simultaneous determination of eight typical fluoroquinolone antibiotics in environmental waters. For that purpose, a broad-specific polyclonal antibody was home-prepared and immobilized on a Sepharose 4B stationary phase for sample affinity extraction achieving an enrichment factor of about 100. Under the optimized conditions, limits of detection and limits of quantitation ranging from 1.2 nM to 5.0 nM and from 4.0 nM to 17 nM, respectively, were obtained. Moreover, very satisfactory recoveries (from 72% to 11%) were reported. In another work, Bailón et al. [18] described the multi-residue determination of penicillins in environmental waters and chicken muscle samples by means of capillary electrophoresis-tandem mass spectrometry using an ion trap analyzer operating in multiple reaction monitoring mode in order to achieve the minimum number of identification points established by the 2002/657/EC European Decision [40]. In contrast to the analysis of chicken muscle where the combination of two SPE procedures using Oasis hydrophilic-lipophilic balance and Alumina N cartridges to obtain satisfactory preconcentration and clean-up of the analyzed samples were required, in the case of environmental waters only the Oasis hydrophilic-lipophilic balance cartridges were necessary to achieve adequate recoveries. With this method, LODs from 0.18 and 0.26 µg/L in water samples were reported.

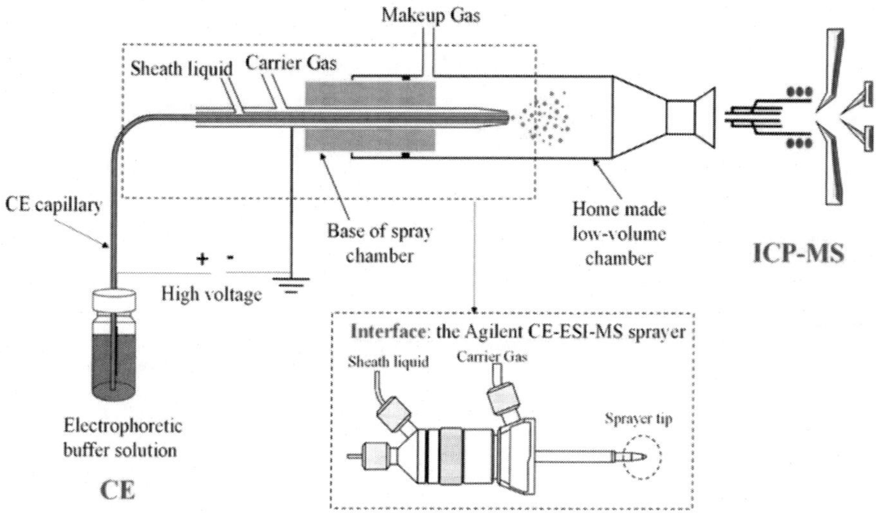

Figure 9. Schematic diagram of the interface for coupling capillary electrophoresis with inductively coupled plasma-mass spectrometry (CE-ICP-MS). Reprinted with permission of reference [76]. Copyright (2014) American Chemical Society.

Capillary electrophoresis coupled with inductively coupled plasma mass spectrometry (CE-ICP-MS) has also play an important role in the element speciation of environmental samples during the last years, and several applications can be found in the literature [74–77]. As an example, Zhao et al. [75] proposed the use of CE-ICP-MS for the ultra-sensitive speciation of mercury in natural waters and fish samples after a microwave-assisted extraction sample treatment procedure. The developed CE-ICP-MS analytical method showed a limit of detection down to 0.021-0.032 ng Hg/mL for MeHg, EtHg and Hg(II) and was proposed for the ultra-trace determination of these mercury species in the commented samples without any preconcentration method after the extraction procedure. Regarding this application field, new interfaces for the CE-ICP-MS are being developed. Recently, Liu et al. [76] developed a simple and highly efficient on-line system coupling capillary electrophoresis to ICP-MS for the simultaneous separation and determination of arsenic and selenium compounds (Figure 9, see references [76, 77] for more details).

The authors coupled the CE to the ICP-MS system by using a sprayer with a novel direct-injection high-efficiency nebulizer chamber as the interface. Using this device, several arsenic and selenium species were baseline separated and determined in a single run within 9 minutes. Limits of detection in the ranges of 0.11-0.37 μg/L and 1.33-2.31 μg/L for arsenic and selenium compounds, respectively, were obtained.

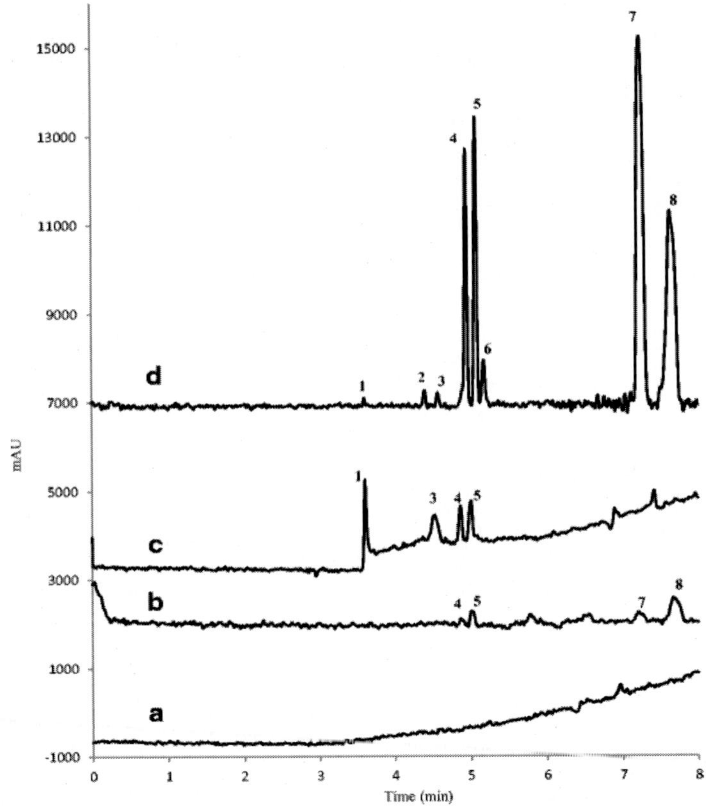

Figure 10. Off-line SPE-FASI-CZE electropherograms of (a) blank river water sample, (b) Barcelona's tap water, (c) Segre River water, and (d) SPE extract of a blank river water sample spiked with benzophenone UV filters at ca. 1 mg/L. UV detection at 345 nm. Peak identification: (1) 2-hydroxy-4-methoxybenzophenone, (2) 2,2'-dihydroxy-4-methoxybenzophenone, (3) 2,2'-dihydroxy-4,4'-dimethoxybenzophenone, (4) 4-hydroxybenzophenone, (5) 2,4-dihydroxybenzophenone, (6) 2,3,4-trihydroxybenzophenone, (7) 4,4'-dihydroxybenzophenone, and (8) 2,2',4,4'-tetrahydroxybenzophenone. Reprinted with permission of reference [79]. Copyright (2014) Springer.

When the employed detection systems are not sensitive enough to reach the required concentration levels for the determination of environmental contaminants according to regulation, enrichment procedures are employed. As commented before, on-line electrophoretic-based preconcentration methods are among the most employed in combination with CZE, also in environmental applications, due to the fact that no special instrumentation requirements are necessary [68]. Strategies based on the stacking phenomena such as LVSS, FASI, and electrokinetic supercharging focusing [78], among others, are the most frequently employed when dealing with CZE separations. Nevertheless, one of the drawbacks of these methodologies when dealing with real environmental samples, especially in the case of environmental waters, is sample salinity. This is due to the fact that sensitivity enhancement with these techniques is typically achieved by the differences in the analyte mobility between the sample region and the running buffer region. In those cases, strategies in order to remove sample salinity or to improve the stacking enhancement effect are required. For example, Liu et al. [70] proposed the use of field-enhanced sample injection in combination with CZE for the determination of aniline and its derivatives in environmental waters. With this approach, limits of detection in the range of 0.29-0.43 ng/mL were obtained. In order to eliminate the effect of the real sample matrix on the stacking efficiency the authors employed the standard addition method for the analysis of water samples from local rivers.

Quesada-Molina et al. [48] described the determination of sulfonylurea herbicides in water and grape samples by CZE using LVSS. In the case of the water samples, an off-line SPE procedure using Oasis HLB cartridges was employed for sample clean-up and preconcentration, being able to achieve limits of detection in the range of 0.04-0.12 μg/L.

In some cases, and especially when the expected concentrations levels to be analyzed are very low, the combination of several preconcentration approaches is a good strategy [69,79]. For example, Purrà et al. [79] proposed the combination of SPE and FASI with CZE for the determination of benzophenone UV filters in environmental water samples. The employment of FASI-CZE allowed the determination of this family of compounds in less than 8 min and achieving a 9- to 25-fold sensitivity enhancement. LODs down to 21-59 μg/L were obtained for most of the targeted benzophenones, with acceptable run-to-run and day-to-day precisions (RSD values lower than 17%). However, the authors observed that sample salinity of environmental water samples was a handicap when FASI approaches need to be employed. In order to remove water sample salinity and to enhance FASI sensitivity, an off-line SPE procedure using a Strata X polymeric reversed-phase sorbent was proposed with recoveries in the range of 72-90% for most of the compounds. As an example, Figure 10 shows the off-line SPE-FASI-CZE electropherograms of (a) a blank river water sample, (b) Barcelona's tap water, (c) Segre River (Spain) water, and (d) SPE extract of a blank river water sample spiked with benzophenones at ca. 1 mg/L.

With the combination of both preconcentration strategies, limits of detection in the range 0.06-0.6 μg/L in a river water matrix were obtained, representing a 2,400- to 6,500-fold sensitivity enhancement. The authors observed that none of the analyzed benzophenones were detected in river water samples collected upstream of industrialized and urban areas, although the presence of some benzophenones, in some cases at relatively high concentrations (10-82 μg/L), was observed in river water samples collected downstream of industrialized and urban areas. The good results obtained in this work show that the proposed off-line SPE-FASI-CZE strategy is suitable for the analysis of benzophenones UV filters in environmental water samples at low microgram per liter levels.

CONCLUSION

Nowadays, capillary zone electrophoresis is becoming very popular in multiple application fields due to its high separation efficiency, rapid separations, and low consumption of reagents as well as of solvents. During the last years CZE has played an important role in food and environmental analysis because the typical sensitivity drawback of this technique when dealing with conventional UV-detection systems is today solved by means of employing more sensitive and selective detectors or by the combination of CZE with off-line and on-line enrichment procedures which are becoming more simple, easy to employ and environment-friendly. In this chapter, an overview of the role of CZE in food and environmental analysis during the early twenty-first century has been presented by describing several applications.

In both, food and environmental analysis CZE has been widely employed for the multi-residue methodologies for instance in the analysis of veterinary drugs, families of organic contaminants, etc. Among the detection systems, electrochemical detection, fluorescence detection or the coupling of capillary electrophoresis to mass spectrometry or tandem mass spectrometry has been described in multiple applications to improve the selectivity and sensitivity on trace analyses approaches. However, even when using these improved detection systems, sometimes the required concentration levels of some specific contaminants, especially in several environmental applications, are too low to be reached. In those cases, the employment of enrichment procedures is required. Despite the great enrichments factors that can be achieved when employing SPE preconcentration procedures (either off-line or on-line), these methods are typically time-consuming. As a consequence, on-line enrichment factors, especially those based on electrophoretic phenomena, are gaining great acceptance in CZE methodologies because they do not need any special instrumentation set-up, and can be easily adapted to any commercially available CE instrument, and can be easily modified (designed) depending on the analytes to be determined and the matrices of the samples to be analyzed. In some cases, the combination of both off-line SPE and on-line electrophoretic-based preconcentration methods are required either to reach very low concentration levels of some specific organic contaminants, or to adapt the sample matrix to improve the sensitive enhancement when, for instance, stacking procedures are employed.

Finally, CZE is playing an important role also in food integrity and authenticity issues. Foodomic profiling either of some specific families of compounds detected by CZE or by employing CZE fingerprints obtained by some universal detectors can be employed as chemical descriptors to address food characterization, classification and authentication by means of chemometric methods.

CZE is today and will continue to be a powerful complementary and alternative methodology to liquid chromatography and gas chromatography in both food and environmental analysis, and sure the number of publications will continue to increase in the future.

Acknowledgments

This work was supported by the Agency for Administration of University and Research Grants (Generalitat de Catalunya, Spain) under the project 2017 SGR-310.

References

[1] Hernández-Borges J, Neusüß C, Cifuentes A, Pelzing M (2004) On-line capillary electrophoresis-mass spectrometry for the analysis of biomolecules. *Electrophoresis* 25:2257–2281. doi: 10.1002/elps.200405954.

[2] Kraly J, Fazal MA, Schoenherr RM, Bonn R, Harwood MM, Turner E, Jones M, Dovichi NJ (2006) Bioanalytical applications of capillary electrophoresis. *Anal Chem* 78:4097–4110. doi: 10.1021/ac060704c.

[3] Siddiqui MR, AlOthman ZA, Rahman N (2017) Analytical techniques in pharmaceutical analysis: A review. *Arab J Chem* 10:S1409–S1421. doi: 10.1016/j.arabjc.2013.04.016.

[4] García-Cañas V, Simó C, Castro-Puyana M, Cifuentes A (2014) Recent advances in the application of capillary electromigration methods for food analysis and Foodomics. *Electrophoresis* 35:147–169. doi: 10.1002/elps.201300315.

[5] Acunha T, Ibáñez C, García-Cañas V, Simó C, Cifuentes A (2016) Recent advances in the application of capillary electromigration

methods for food analysis and Foodomics. *Electrophoresis* 37:111–141. doi: 10.1002/elps.201500291.

[6] Piñero MY, Bauza R, Arce L (2011) Thirty years of capillary electrophoresis in food analysis laboratories: Potential applications. *Electrophoresis* 32:1379–1393. doi: 10.1002/elps.201000541.

[7] Vallejo-Cordoba B, González-Córdova AF (2010) Capillary electrophoresis for the analysis of contaminants in emerging food safety issues and food traceability. *Electrophoresis* 31:2154–2164. doi: 10.1002/elps.200900777.

[8] Hamdan II (2017) Capillary electrophoresis in the analysis of pharmaceuticals in environmental water: a critical review. *J Liq Chromatogr Relat Technol* 40:111–125. doi: 10.1080/10826076.2017.1293550.

[9] Juan-García A, Font G, Picó Y (2005) Determination of organic contaminants in food by capillary electrophoresis. *J Sep Sci* 28:793–812. doi: 10.1002/jssc.200500041.

[10] Deng Y, Gasilova N, Qiao L, Zhou YL, Zhang XX, Girault HH (2014) Highly sensitive detection of five typical fluoroquinolones in low-fat milk by field-enhanced sample injection-based CE in bubble cell capillary. *Electrophoresis* 35:3355–3362. doi: 10.1002/elps.201400294.

[11] Chen S, Xu Y, Bi Y, Du W, Liu BF (2006) Analysis of environmental pollutants metabolized from pesticides using capillary electrophoresis with multiphoton-excited fluorescence detection. *Talanta* 70:63–67. doi: 10.1016/j.talanta.2005.11.044.

[12] Ji H, Wu Y, Duan Z, Yang F, Yuan H, Xiao D (2017) Sensitive determination of sulfonamides in environmental water by capillary electrophoresis coupled with both silvering detection window and in-capillary optical fiber light-emitting diode-induced fluorescence detector. *Electrophoresis* 38:452–459. doi: 10.1002/elps.201600364.

[13] Ferey L, Delaunay N, Rutledge DN, Huertas A, Raoul Y, Gareil P, Vial J (2013) Use of response surface methodology to optimize the simultaneous separation of eight polycyclic aromatic hydrocarbons by capillary zone electrophoresis with laser-induced fluorescence

detection. *J Chromatogr A* 1302:181–190. doi: 10.1016/j.chroma. 2013.06.027.

[14] Chu Q, Wang J, Zhang D, Ye J (2010) Sensitive determination of parabens in soy sauces by capillary zone electrophoresis with amperometric detection. *Eur Food Res Technol* 231:891–897. doi: 10.1007/s00217-010-1343-2.

[15] Chu Q, Zhang D, Wang J, Ye J (2009) Multi-residue analysis of sulfonamides in animal tissues by capillary zone electrophoresis with electrochemical detection. *J Sci Food Agric* 89:2498–2504. doi: 10.1002/jsfa.3756.

[16] Himmelsbach M, Klampfl CW, Buchberger W (2005) Development of an analytical method for the determination of antidepressants in water samples by capillary electrophoresis with electrospray ionization mass spectrometric detection. *J Sep Sci* 28:1735–1741. doi: 10.1002/jssc.200500157.

[17] Lara FJ, García-Campaña AM, Alés-Barrero F, Bosque-Sendra JM, García-Ayuso LE (2006) Multiresidue method for the determination of quinolone antibiotics in bovine raw milk by capillary electrophoresis-tandem mass spectrometry. *Anal Chem* 78:7665–7673. doi: 10.1021/ac061006v.

[18] Bailón-Pérez MI, García-Campaña AM, del Olmo-Iruela M, Cruces-Blanco C, Gracia LG (2009) Multiresidue determination of penicillins in environmental waters and chicken muscle samples by means of capillary electrophoresis-tandem mass spectrometry. *Electrophoresis* 30:1708–1717. doi: 10.1002/elps.200800732.

[19] Rodríguez-Gonzalo E, Domínguez-Álvarez J, García-Gómez D, García-Jiménez MG, Carabias-Martínez R (2010) Determination of endocrine disruptors in honey by CZE-MS using restricted access materials for matrix cleanup. *Electrophoresis* 31:2279–2288. doi: 10.1002/elps.200900715.

[20] Tejada-Casado C, Moreno-González D, Lara FJ, García-Campaña AM, del Olmo-Iruela M (2017) Determination of benzimidazoles in meat samples by capillary zone electrophoresis tandem mass

spectrometry following dispersive liquid–liquid microextraction. *J Chromatogr A* 1490:212–219. doi: 10.1016/j.chroma.2017.02.023.

[21] Pedersen-Bjergaard S, Rasmussen KE, Halvorsen TG (2000) Liquid-liquid extraction procedures for sample enrichment in capillary zone electrophoresis. *J Chromatogr A* 902:91–105. doi: 10.1016/S0021-9673(00)00738-X.

[22] Verónica C. S, María De Los Ángeles F, Claudio R. G, Fernanda SM (2014) Analysis of phenolic compounds in onion nectar by miniaturized off-line solid phase extraction-capillary zone electrophoresis. *Anal Methods* 6:4878–4884. doi: 10.1039/c4ay00240g.

[23] Kitagawa F, Otsuka K (2014) Recent applications of on-line sample preconcentration techniques in capillary electrophoresis. *J Chromatogr A* 1335:43–60. doi: 10.1016/j.chroma.2013.10.066.

[24] Núñez O (2015) On-line electrophoretic-based preconcentration methods in capillary zone electrophoresis: principles and relevant applications. In: Reed, C (Ed.) *Capillary Electrophoresis* (CE), Principles, Challenges and Applications, Nova Science Publishers, Inc., New York, USA, pp. 73-124.

[25] Castañeda G, Rodríguez-Flores J, Ríos A (2005) Analytical approaches to expanding the use of capillary electrophoresis in routine food analysis. *J Sep Sci* 28:915–924. doi: 10.1002/jssc.200500036.

[26] Cifuentes A (2006) Recent advances in the application of capillary electromigration methods for food analysis. *Electrophoresis* 27:283–303. doi: 10.1002/elps.200500474.

[27] García-Cañas V, Cifuentes A (2008) Recent advances in the application of capillary electromigration methods for food analysis. *Electrophoresis* 29:294–309. doi: 10.1002/elps.200700438.

[28] Bermudo E, Núñez O, Puignou L, Galceran MT (2006) Analysis of acrylamide in food samples by capillary zone electrophoresis. *J Chromatogr A* 1120:199–204. doi: 10.1016/j.chroma.2005.10.074.

[29] Bermudo E, Núñez O, Puignou L, Galceran MT (2006) Analysis of acrylamide in food products by in-line preconcentration capillary zone

electrophoresis. *J Chromatogr A* 1129:129–134. doi: 10.1016/j.chroma.2006.06.076.

[30] Casado-Terrones S, Segura-Carretero A, Busi S, Dinelli G, Fernández-Gutiérrez A (2007) Determination of tetracycline residues in honey by CZE with ultraviolet absorbance detection. *Electrophoresis* 28:2882–2887. doi: 10.1002/elps.200600614.

[31] Bailón-Pérez MI, García-Campaña AM, Cruces-Blanco C, Del Olmo Iruela M (2007) Large-volume sample stacking for the analysis of seven beta-lactam antibiotics in milk samples of different origins by CZE. *Electrophoresis* 28:4082–90. doi: 10.1002/elps.200700305.

[32] Soto-Chinchilla JJ, García-Campaña AM, Gámiz-Gracia L (2007) Analytical methods for multiresidue determination of sulfonamides and trimethoprim in meat and ground water samples by CE-MS and CE-MS/MS. *Electrophoresis* 28:4164–4172. doi: 10.1002/elps.200600848.

[33] Li T, Shi ZG, Zheng MM, Feng YQ (2008) Multiresidue determination of sulfonamides in chicken meat by polymer monolith microextraction and capillary zone electrophoresis with field-amplified sample stacking. *J Chromatogr A* 1205:163–170. doi: 10.1016/j.chroma.2008.08.017.

[34] He HB, Lv XX, Yu QW, Feng YQ (2010) Multiresidue determination of (fluoro)quinolone antibiotics in chicken by polymer monolith microextraction and field-amplified sample stacking procedures coupled to CE-UV. *Talanta* 82:1562–1570. doi: 10.1016/j.talanta.2010.07.055.

[35] Martínez-Villalba A, Núñez O, Moyano E, Galceran MT (2013) Field amplified sample injection-capillary zone electrophoresis for the analysis of amprolium in eggs. *Electrophoresis* 34:870–876. doi: 10.1002/elps.201200579.

[36] Zhao S, Zhao H, Zhu H, Li C, Lu X, Bi S (2014) Determination of norfloxacin and ciprofloxacin in chicken meat based on matrix solid-phase dispersion extraction and capillary zone electrophoresis. *J Chem*. doi: 10.1155/2014/610120.

[37] García-Campaña AM, Gámiz-Gracia L, Lara FJ, Del Olmo Iruela M, Cruces-Blanco C (2009) Applications of capillary electrophoresis to the determination of antibiotics in food and environmental samples. *Anal Bioanal Chem* 395:967–986. doi: 10.1007/s00216-009-2867-9.

[38] Flurer CL (2001) Analysis of antibiotics by capillary electrophoresis. *Electrophoresis* 22:4249–4261. doi: 10.1002/1522-2683(200111) 22:19<4249::AID-ELPS4249>3.0.CO;2-8.

[39] Malá Z, Šlampová A, Gebauer P, Boček P (2009) Contemporary sample stacking in CE. *Electrophoresis* 30:215–229. doi: 10.1002/elps.200800433.

[40] Commission Decision 2002/657/EC of 12 August 2001 implementing Council Directive 96/23/EC cocerning the performance of analytical methods and the interpretation of results, *Official Journal of the European Communities* L221 8-36.

[41] Yan N, Zhou L, Zhu Z, Chen X (2009) Determination of melamine in dairy products, fish feed, and fish by capillary zone electrophoresis with diode array detection. *J Agric Food Chem* 57:807–811. doi: 10.1021/jf803429e.

[42] Xia J, Zhou N, Liu Y, Chen B, Wu Y, Yao S (2010) Simultaneous determination of melamine and related compounds by capillary zone electrophoresis. *Food Control* 21:912–918. doi: 10.1016/j.foodcont. 2009.12.009.

[43] Kong Y, Wei C, Wang Z, Hou Z, Li H, Yu J, Yuan J (2014) Determination of Melamine in Quail Egg and Milk with Capillary Zone Electrophoresis. *Am J Anal Chem* 5:258–265.

[44] Rodriguez R, Boyer I, Font G, Pico Y (2001) Capillary zone electrophoresis for the determination of thiabendazole, prochloraz and procymidone in grapes. *Analyst* 126:2134–2138. doi: 10.1039/ b106478a.

[45] Aulakh JS, Fekete A, Malik AK, Schmitt-Kopplin P, Mahajan RK (2007) Determination of Quats in Beverages and Urine Samples by Capillary Zone Electrophoresis. *Ann Chim* 97:1157–1167.

[46] Lin X, Hong Q, Wu X, Guo L, Xie Z (2008) Analysis of phenoxyl-type N-methylcarbamate pesticide residues in vegetables by capillary

zone electrophoresis with pre-column hydrolysis and amperometric detection. *J Chromatogr Sci* 46:615–621. doi: 10.1093/chromsci/46.7.615.

[47] Moreno M, Bermejo E, Sánchez A, Chicharro M, Zapardiel A (2008) Application of matrix solid-phase dispersion to the determination of amitrole and urazole residues in apples by capillary electrophoresis with electrochemical detection. *Anal Bioanal Chem* 391:867–872. doi: 10.1007/s00216-008-2058-0.

[48] Quesada-Molina C, Del Olmo-Iruela M, García-Campaña AM (2010) Trace determination of sulfonylurea herbicides in water and grape samples by capillary zone electrophoresis using large volume sample stacking. *Anal Bioanal Chem* 397:2593–2601. doi: 10.1007/s00216-010-3812-7.

[49] Miller JD, Trenhorn HL, Mycotoxins in Grain. *Compounds Other than Aflatoxin*, 2nd Edn., Eagan Press, St. Paul, MN 1994.

[50] Maragos CM (2004) Detection of moniliformin in maize using capillary zone electrophoresis. *Food Addit Contam* 21:803–810. doi: 10.1080/02652030410001717812.

[51] Arribas AS, Bermejo E, Zapardiel A, Téllez H, Rodríguez-Flores J, Zougagh M, Ríos Á, Chicharro M (2009) Screening and confirmatory methods for the analysis of macrocyclic lactone mycotoxins by CE with amperometric detection. *Electrophoresis* 30:499–506. doi: 10.1002/elps.200800305.

[52] Wenzl T, Lachenmeier DW, Gökmen V (2007) Analysis of heat-induced contaminants (acrylamide, chloropropanols and furan) in carbohydrate-rich food. *Anal Bioanal Chem* 389:119–137. doi: 10.1007/s00216-007-1459-9.

[53] Fei XQ, Li C, Yu XD, Chen HY (2007) Determination of heterocyclic amines by capillary electrophoresis with UV-DAD detection using on-line preconcentration. *J Chromatogr B Anal Technol Biomed Life Sci* 854:224–229. doi: 10.1016/j.jchromb.2007.04.031.

[54] Delgado-Andrade C, Rufián-Henares JA, Morales FJ (2005) Fast method to determine furosine in breakfast cereals by capillary zone

electrophoresis. *Eur Food Res Technol* 221:707–711. doi: 10.1007/s00217-005-0030-1.

[55] Albillos SM, Busto MD, Perez-Mateos M, Ortega N (2005) Chemometrical analysis of capillary electrophoresis casein fractions for predicting ripening times of milk mixture cheese. *J Agric Food Chem* 53:6094–6099. doi: 10.1021/jf050352v.

[56] Andersen KE, Bjergegaard C, Møller P, Sørensen JC, Sørensen H (2005) Compositional variations for α-galactosides in different species of Leguminosae, Brassicaceae, and barley: A chemotaxonomic study based on chemometrics and high-performance capillary electrophoresis. *J Agric Food Chem* 53:5809–5817. doi: 10.1021/jf040471v.

[57] McCourt J, Stroka J, Anklam E (2005) Experimental design-based development and single laboratory validation of a capillary zone electrophoresis method for the determination of the artificial sweetener sucralose in food matrices. *Anal Bioanal Chem* 382:1269–1278. doi: 10.1007/s00216-005-3258-5.

[58] Kvasnička F (2005) Capillary electrophoresis in food authenticity. *J Sep Sci* 28:813–825. doi: 10.1002/jssc.200500054.

[59] Vallejo-Cordoba B, González-Córdoba A, Mazorra-Manzano MA, Rodríguez-Ramírez R (2005) Capillary eletrophoresis for the analysis of meat authenticity. *J Sep Sci* 28:826–836. doi: 10.1002/jssc.200500013.

[60] Franquet-Griell H, Checa A, Núñez O, Saurina J, Hernández-Cassou S, Puignou L (2012) Determination of polyphenols in spanish wines by capillary zone electrophoresis. Application to wine characterization by using chemometrics. *J Agric Food Chem* 60:8340–8349. doi: 10.1021/jf302078j.

[61] Navarro-Pascual-Ahuir M, Lerma-Garcia MJ, Simo-Alfonso EF, Herrero-Martinez JM (2015) Rapid differentiation of commercial juices and blends by using sugar profiles obtained by capillary zone electrophoresis with indirect UV detection. *J Agric Food Chem* 63:2639–2646. doi: 10.1021/acs.jafc.5b00122.

[62] Navarro-Pascual-Ahuir M, Lerma-García MJ, Simó-Alfonso EF, Herrero-Martínez JM (2015) Quality control of fruit juices by using organic acids determined by capillary zone electrophoresis with poly(vinyl alcohol)-coated bubble cell capillaries. *Food Chem* 188:596–603. doi: 10.1016/j.foodchem.2015.05.057.

[63] de Oliveira Mendes T, Porto BLS, Bell MJV, Perrone ÍT, de Oliveira MAL (2016) Capillary zone electrophoresis for fatty acids with chemometrics for the determination of milk adulteration by whey addition. *Food Chem* 213:647–653. doi: 10.1016/j.foodchem.2016.07.035.

[64] Duarte LM, Adriano LHC, de Oliveira MAL (2018) Capillary electrophoresis in association with chemometrics approach for bitterness hop (Humulus lupulus L.) classification. *Electrophoresis* 39:1399–1409. doi: 10.1002/elps.201700420.

[65] Navarro M, Núñez O, Saurina J, Hernández-Cassou S, Puignou L (2014) Characterization of fruit products by capillary zone electrophoresis and liquid chromatography using the compositional profiles of polyphenols: Application to authentication of natural extracts. *J Agric Food Chem* 62:1038–1046. doi: 10.1021/jf404776d.

[66] Timerbaev AR, Fukushi K (2003) Analysis of seawater and different highly saline natural waters by capillary zone electrophoresis. *Mar Chem* 82:221–238. doi: 10.1016/S0304-4203(03)00071-9.

[67] Paull B, King M (2003) Quantitative capillary zone electrophoresis of inorganic anions. *Electrophoresis* 24:1892–1934. doi: 10.1002/elps.200305410.

[68] Núñez O (2015) Environmental Applications of Capillary Zone Electrophoresis, in Barbooti, M (Ed.) *Environmental Applications of Instrumental Chemical Analysis*, CRC Press, Wareton NJ, USA, pp. 305-334.

[69] Quesada-Molina C, Olmo-Iruela M del, García-Campaña AM (2012) Analysis of cephalosporin residues in environmental waters by capillary zone electrophoresis with off-line and on-line preconcentration. *Anal Methods* 4:2341. doi: 10.1039/c2ay25179e.

[70] Liu S, Wang W, Chen J, Sun J (2012) Determination of aniline and its derivatives in environmental water by capillary electrophoresis with on-line concentration. *Int J Mol Sci* 13:6863–6872. doi: 10.3390/ijms13066863.

[71] Robledo VR, Smyth WF (2009) The application of CE-MS in the trace analysis of environmental pollutants and food contaminants. *Electrophoresis* 30:1647–1660. doi: 10.1002/elps.200800651.

[72] Safarpour H, Asiaie R, Katz S (2004) Quantitative analysis of imazamox herbicide in environmental water samples by capillary electrophoresis electrospray ionization mass spectrometry. *J Chromatogr A* 1036:217–222. doi: 10.1016/j.chroma.2004.03.002.

[73] Zhang XH, Deng Y, Zhao MZ, Zhou YL, Zhang XX (2018) Highly-sensitive detection of eight typical fluoroquinolone antibiotics by capillary electrophoresis-mass spectroscopy coupled with immunoaffinity extraction. *RSC Adv* 8:4063–4071. doi: 10.1039/c7ra12557g.

[74] Sun J, He B, Yin Y, Li L, Jiang G (2010) Speciation of organotin compounds in environmental samples with semi-permanent coated capillaries by capillary electrophoresis coupled with inductively coupled plasma mass spectrometry. *Anal Methods* 2:2025–2031. doi: 10.1039/c0ay00558d.

[75] Zhao Y, Zheng J, Fang L, Lin Q, Wu Y, Xue Z, Fu F (2012) Speciation analysis of mercury in natural water and fish samples by using capillary electrophoresis-inductively coupled plasma mass spectrometry. *Talanta* 89:280–285. doi: 10.1016/j.talanta.2011.12.029.

[76] Liu L, Yun Z, He B, Jiang G (2014) Efficient interface for online coupling of capillary electrophoresis with inductively coupled plasma-mass spectrometry and its application in simultaneous speciation analysis of arsenic and selenium. *Anal Chem* 86:8167–8175. doi: 10.1021/ac501347d.

[77] Liu L, He B, Yun Z, Sun J, Jiang G (2013) Speciation analysis of arsenic compounds by capillary electrophoresis on-line coupled with

inductively coupled plasma mass spectrometry using a novel interface. *J Chromatogr A* 1304:227–233. doi: 10.1016/j.chroma.2013.07.034.

[78] Botello I, Borrull F, Aguilar C, Calull M (2010) Electrokinetic supercharging focusing in capillary zone electrophoresis of weakly ionizable analytes in environmental and biological samples. *Electrophoresis* 31:2964–2973. doi: 10.1002/elps.201000184.

[79] Purrà M, Cinca R, Legaz J, Núñez O (2014) Solid-phase extraction and field-amplified sample injection-capillary zone electrophoresis for the analysis of benzophenone UV filters in environmental water samples. *Anal Bioanal Chem* 406:6189–6202. doi: 10.1007/s00216-014-8086-z.

In: Capillary Electrophoresis ... ISBN: 978-1-53615-223-4
Editors: O. Núñez and X. Subirats

Chapter 2

THE ROLE OF CAPILLARY ELECTROPHORESIS IN THE CHEMOMETRIC CHARACTERIZATION OF FOOD PRODUCTS

Sonia Sentellas and Javier Saurina*
Department of Chemical Engineering and Analytical Chemistry,
University of Barcelona, Barcelona, Spain

ABSTRACT

The characterization of food products based on compositional profiles of natural components obtained by capillary electrophoresis (CE) is discussed here. CE with different detection techniques, especially UV-visible spectroscopy and mass spectrometry, has widely been used for the generation of data to be further processed by chemometrics. It has been demonstrated that contents of food constituents are dependent on climatic, agricultural and manufacturing issues so they can be exploited as chemical descriptors of food features and quality. Chemometrics has widely been used as a way of facilitating the extraction of information dealing with food properties. In this regard, cluster analysis, principal component analysis and related methods are currently used for discrimination, classification,

* Corresponding Author's E-mail: xavi.saurina@ub.edu.

modelling and authentication purposes. We will comment on some representative examples to illustrate the possibilities of CE combined with chemometrics for the study of food products.

Keywords: chemometrics, profiling, fingerprinting, food characterization, food classification, food authentication

INTRODUCTION

Our society is increasingly concerned with the quality and authenticity of food products because of several reasons including economical, organoleptic, nutritional, healthy or toxicological issues [1-3]. A large number of analytical methods have been published for the characterization of food products such as wine, beer, honey, tea, olive oil, fruits and juices, vegetables and dairy products in order to assess some important features such as geographical origin, species and variety, elaboration practices, etc. Studies typically rely on high performance separation techniques combined with spectroscopic or mass spectrometric (MS) detection [4-6].

Capillary electrophoresis (CE) is among the most promising techniques for the analysis of food products [7]. CE provides some analytical advantages over other instrumental platforms regarding high separation performance, fast runs, low consumption of samples and reagents. As a result, this technique suitable for the analysis of large series of samples in the course of characterization, classification and authentication studies [7, 8]. The low sensitivity and poor detection limits of CE methods result in a serious drawback to tackle the determination of trace components in complex food samples. In the last years, however, significant improvements have been achieved for enhancing the sensitivity of CE methods based on innovative derivatization strategies and on-line sample preconcentration [9].

Several families of natural food components, including organic acids, alcohols, esters, amino acids, biogenic amines, inorganic species and phenolic compounds have been exploited as potential descriptors of food products [10]. Contents of some of these compounds may vary depending on the sample characteristics so both concentrations or related instrumental responses can be used as the data for exploratory studies from which selective chemical markers or predominant components can be encountered.

Beyond descriptive issues, more specific classification models can also be established to assign unknown samples to predefined categories [11]. Furthermore, food authentication is another aspect stirring up the interest of the scientific community and consumers because of its economic and social implications. In authentication, the analytical issue deals with the identification/assignation of a given sample as genuine or fraud according to levels of some markers. For example, in oil industry, extra virgin olive oils (EVOO) from a given Protected Designation of Origin (PDO) may be adulterated with a percentage of another oil of lower quality such as refined olive oil or sunflower oil. In that case, a main objective will be the quantification of percentage of adulterant oils in the unknown samples under study [12]. Another illustrative case concerns the evaluation of cranberry-based nutraceuticals that should be elaborated from American red cranberry (*Vaccinium macrocarpon*). Commercial products may sometimes contain variable amounts of other cranberry species which may lack of the desired bioactivity; in more extreme cases, products can be adulterated with cheaper fruits such as grape or raspberry. The type of products or ingredients used in these fraudulent practices can tentatively be deduced from the trace of components coming from the adulterant [13]. Amounts or percentages of adulteration can be estimated as well.

The treatment of the huge amount of CE data resulting from the analysis of large series of food samples entails some difficulties in order to extract the relevant information. Univariate statistics can help to reveal some sample patterns. For instance, boxplots are useful to compare levels of compounds among different groups/classes of samples. However, chemometric methods for multivariate analysis seem to be much more recommendable for a comprehensive assessment of sample characteristics. In this case, experimental variables, i.e., compositional profiles or instrumental fingerprints, are processed and overall information can be gained from Principal Component Analysis (PCA) and related methods. As a result, patterns and relationships among samples and variables can easily be deduced [1, 2, 10]. Multivariate methods can also be applied to classification and quantification.

Some important aspects arising from the combination of CE and chemometrics will be described in the following sections. Data types, data pretreatments and chemometric methods will be first discussed. In another section, illustrative cases tackling the application of CE to food analysis will be commented. Finally, remarks and further trends will be presented in the "Conclusions" section.

DATA ANALYSIS

Electrophoretic Data

Electrophoretic techniques provide data of different complexity that can be exploited for characterization, classification and authentication purposes mainly using chemometric methods of analysis. As detailed below, the main approaches considered here to deal with these objectives rely on (*i*) concentrations of some chemical species (profiling) and (*ii*) instrumental – absorbance or intensity- features (fingerprinting) [11]. Hence, when several samples are analyzed simultaneously, the resulting data is arranged in a matrix, where each row corresponds to a given sample and each column to a compound/feature. Accordingly, dimensions of this matrix are $m \times n$, being m the number of samples and n the number of compounds/features.

Profiling is based on targeted analysis in which the concentrations of selected species are used as the source of information. Hence, compounds of interest are previously quantified and the resulting data is used to build the matrix (see scheme in Figure 1a). As the main disadvantage of this approach, when dealing with complex food products the quantification of these compounds may result in a difficult task due to the concurrence of dozens of compounds, some of them even unknown, that may interfere with by the determination.

Fingerprinting strategies perform untargeted analysis of instrumental responses without assuming any previous knowledge. Sample fingerprints can be obtained from different instrumental means such as infrared spectroscopy, nuclear magnetic resonance (NMR), liquid and gas chromatography, MS and LC-MS and, of course, CE. In the specific case of CE, electropherograms are recorded typically by UV-vis absorption, fluorescence and MS. As a result, each sample under analysis provides a list of absorbance, fluorescence, or mass-to-charge (*m/z*) values recorded as a function of time (see scheme in Figure 1b). These peak signals are commonly referred to as features. When dealing with multi-way detectors, such as diode array spectrophotometers or mass spectrometers, spectral information is also available throughout the entire electropherogram so that richer spectro/electrophoretic fingerprints can be obtained. Compared with the profiling approach, prior identification and quantification of analytes is not required anymore. Interestingly, information from both well-known and unknown components can be gained since instrumental profiles reflect the complexity of the sample and, although components are not explicitly quantified, signal intensities obviously depend on the concentrations of all constituents. Nowadays, fingerprint analysis is gaining popularity because of its simplicity. However, from the point of view of food science, the identification of specific markers responsible for product characteristics is greatly welcome.

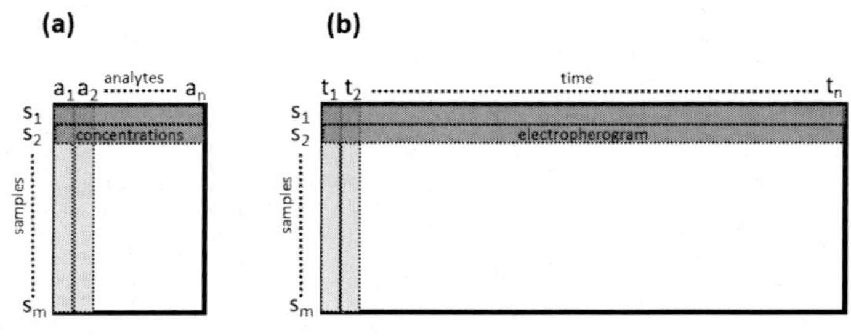

Figure 1. Matrix arrangements corresponding to (a) profiling and (b) fingerprinting approaches.

Data Preprocessing

CE data is often preprocessed to improve its quality and provide better sample descriptors before being subjected to a comprehensive chemometric analysis. In profiling, the data treatment is, in general, quite simple. If necessary, mean-centering or autoscaling is performed in order to equalize the contribution of both major and minor sample constituents on the model.

In fingerprinting, procedures to polish CE data are much more varied. Time and baseline drifts sometimes occur in the electropherograms and corrective procedures may be applied. For instance, fluctuations in the migration time of analytes result in an important source of variability among runs that may affect the model performance, especially when dealing with large series of analyses. Peak shifting can be minimized via alignment procedures in which the position of one (or various) reference peak(s) is(are) used for synchronization. Complementary peak width correction may also be needed for peak broadening control. Background spectral correction by subtracting the baseline spectrum before the peak appearance is another treatment often applied. The drift in the electropherogram baseline can be circumvented by absorbance subtraction or baseline modeling. Other sophisticated treatments for detrending and noise filtering can be applied to electropherograms such as moving–window average and Savitzky–Golay filters, wavelets and artificial neural networks [10, 11].

Chemometric Methods

Multivariate methods such as Principal Component Analysis (PCA), Partial Least Squares (PLS) regression and Partial Least Squares-Discriminant Analysis (PLS-DA) are typically applied to sample characterization, quantification and classification [10]. Some comments of these methods are given as follows.

PCA is the most widely used method for exploratory studies in food characterization. PCA relies on the concentration of the information contained in the original variables (i.e., concentrations of selected species or instrumental fingerprints) into a small number of new mathematical variables, the so-called principal components (PCs). The data matrix is decomposed into two smaller matrices of scores (coordinates of the samples) and loadings (eigenvalues), providing information on samples and variables, respectively. The first principal component (PC1) is calculated to retain the maximum amount of data variance from the whole data set; the following PC, PC2, retains the maximum amount of the unexplained (residual) variance and is orthogonal to PC1. Further PCs are calculated in the same way until all the relevant information is captured by the model. As a result, a high percentage of chemical information can be displayed efficiently from scores' and loadings' plots of the first PCs. The scatter plot of scores represents the map of distribution of samples, thus revealing patterns, similarities and differences that might be attributed to food product properties. For instance, characteristics such as origin, manufacturing, product varieties and so on can be inferred from the study of the scores' plot. Complementarily, the behavior of variables (here analyte concentrations or instrumental fingerprints) can be studied from the plots of loadings. Hence, variables with high descriptive or discriminant potential can be identified and selected. The scatter plot of loadings results in an excellent way to study the correlation among variables. Besides, relationships between samples and variables can also be investigated from the simultaneous study of scores and loadings via the so-called bi-plot.

PLS is typically applied to the determination of concentrations of analytes, impurities and/or adulteration percentages. The algorithm takes into account the X-data matrix of responses (either from profiling or fingerprinting) and the Y-data matrix of concentrations to build a multivariate calibration model using several well-known (genuine and adulterated) samples as the standards. Mathematical factors, here referred to as latent variables (LVs), are then calculated to maximize the correlation between X- and Y-data matrices. The optimum number of LVs to be included in the model is commonly established by cross-validation using leave-one-out or Venetian blind approaches. In a second step, once the model is build and validated, it can be used to predict analyte/

impurity/adulteration concentrations on a new set of unknown food samples with similar characteristics.

PLS-DA is an excellent tool to classify food samples into predefined categories. As in the case of PLS, this method uses standard samples belonging to well-known classes to stablish a mathematical relationship between responses (X-data) and sample classes (Y-data). Y-data is coded numerically (e.g., 1, 2, 3, …) to express the belonging of each sample to its class. Again, the assessment of the model requires the estimation of the proper number of LVs to be used. Subsequently, the model can be applied to assign test samples into their corresponding classes. Apart from PLS-DA, cluster analysis (CA) and Soft Independent Modelling of Class Analogies (SIMCA) are sometimes used in classification studies. CA is an unsupervised pattern recognition method to group objects on the basis of their similarities or differences. Results are graphically presented as a classification tree or dendrogram in which samples are linked according to a distance measurement among groups (e.g., Euclidean, squared-Euclidean distances). SIMCA accounts the probability of each sample to belong to defined classes based on statistic parameters such as Q and T2 derived from Chi-squared and T-student tests, respectively. Models of probability of classification at desired significance levels can thus be established.

After the application of mathematical methods, chemometric results need to be thoroughly interpreted to gain meaningful information. In profiling, chemical markers consisting of compounds up- or down-expressed depending on the classes can be deduced. In the most favorable situations, specific chemical components of a given class can be identified. On the contrary, in fingerprinting, chemical compounds responsible for the instrumental output are frequently unknown. At this stage, the identification of potential descriptors is, by far, the most difficult tasks. In this regard, unknown compounds can be elucidated by MS, MS/MS and/or NMR assays. Finally, those compounds that have tentatively been assigned can be confirmed by comparison of experimental data with that compiled in databases of exact m/z values, MS/MS and NMR spectra and, if available, with the corresponding chemical standards.

APPLICATIONS TO FOOD ANALYSIS

Table 1 summarizes some recent applications of CE combined with chemometrics to food characterization, classification and authentication. The most widely used electrophoretic modes were capillary zone electrophoresis (CZE) and micellar electrokinetic chromatography (MEKC). As it can be seen, fresh fruit and fruit derivatives, dairy products (e.g., milk and cheese), wine, tea, and water were some popular matrices successfully analyzed in this way. Regarding the analytes, amino acids, peptides, organic acids, (poly)phenolic compounds, sugars and inorganic ions were considered as descriptors of interest. Some selected applications are commented in detail below.

In the example by Navarro and coworkers, fruit-based natural extracts were characterized and authenticated on the basis of polyphenolic compounds determined by CZE with UV-vis detection. The separation was performed with fused-silica capillaries of 75 μm internal diameter (ID) using 35 mM sodium tetraborate (pH 9.4) with 5% methanol as background electrolyte (BGE) solution. Several fresh fruits (cranberry, grapes, blueberry and raspberry), fruit-based products and commercial cranberry-based nutraceuticals with antibiotic activity were analyzed. The sample treatment consisted of solvent extraction with acidified hydro-organic solutions combined with extract purification by size-exclusion chromatography using a polysaccharide resin (Sephadex LC-20) to isolate the active proanthocyanidins (PACs). PCA models from electrophoretic data were able to discriminate samples according to the fruit of origin (see Figure 2). Authors found a suspicious cranberry sample, possibly adulterated or wrongly assigned, as it appeared in the area of predominance of grape samples. Further analysis of this sample by high resolution mass spectrometry (HRMS) showed that it did not contain the claimed active A-type PACs, highly characteristic of cranberry-based products, so its authenticity was doubtful [23].

Table 1. Recent applications of CE combined with chemometrics to food characterization, classification and authentication

Matrix	Analytes	Separation technique	Detection	Chemometric method	Comments	Ref.
Water	Anions (chloride, sulfate, nitrate and fluoride)	CZE		Hierarchical cluster analysis and PCA	Water quality in the Golcuk-Isparta region and classification of resources	[14]
Fruit juices	Organic acids	CZE	Indirect UV	LDA	Differentiation of juices according to fruit type	[15]
Milk	Fatty acids	CZE		Correlation coefficient	Milk adulteration by whey addition	[16]
Passion fruit juice	Aminoacids	MEKC	UV	PCA	Detection of adulteration	[17]
Green tea	Catechins	MEKC		Pattern recognition methods, PCA, hierarchical cluster analysis, LDA and quadratic DA	Determination of origin	[18]
Fruit juices	Organic acids	CZE	UV-vis ("bubble cell" capillary)	LDA and MLR	Differentiation of juices according to fruit type and prediction of percentage of grapefruit juice added to pineapple and orange juices	[19]
Fruit juices	Sugar and cyclamate	CZE	Indirect UV	LDA and MLR	Differentiation of fruit type and determination of adulteration with grape juice	[20]
Green tea leaves	Catechins, methylxanthine theobromine and caffeine	CD-MEKC			Chemometric methods applied after electronic nose analysis	[21]
White wines	Polyphenols	CZE	Amperometric	PCA, discriminant analysis	Classification by grape variety	[22]

Matrix	Analytes	Separation technique	Detection	Chemometric method	Comments	Ref.
Fruit juices, nutraceuticals	Polyphenols	CZE	UV & MS	PCA	Classification and authentication of cranberry based juices and nutraceuticals	[23]
Avocado fruits	Phenolic acids	CZE	UV	PCA	Differences in variety and ripening stages	[24]
Wine	Polyphenols	CZE		PCA	Discrimination according to origin	[25]
Olive oil	Proteins	CGE	MS	Multivariate tools	Discrimination according to variety and origin	[26]
Cheese	Casein	CZE	UV	PLS-1, PLS-2, PCR and MLR	Type of milk (bovine, ovine and caprine) content	[27]
Cheese	Caseins	CZE	UV	PLS, PCR and MLR	Bovine, ovine and caprine milk content determination	[28]
Wine	Biogenic amine	CZE	UV	PCA	Wine characterization according to origin and aging	[29]
Wine	Polyphenols	CE microchip	Electrochemical	PCA, LDA and CART	Classification	[30]
Cheese	Casein	CZE	UV	PLS, PCR	Prediction of ripening time	[31]
Red wines	Metals (Na, K, Mg, Mn, Li)	CZE		Cluster analysis, PCA, LDA, KNN, SIMCA	Categorization according the origin	[32]
Milk	Protein	CZE		SIMCA	Milk adulteration	[33]

PCA: Principal Component Analysis; LDA: Linear Discriminant Analysis; DA: Discriminant Analysis; MLR: Multiple Linear Regression; PLS: Partial Least Squares; PCR: Principal Component Regression; CART: Classification and Regression Tree; KNN: K-nearest neighbors; SIMCA: Soft Independent Modelling of Class Analogies.

The characterization of passion fruit juices with PCA was based on compositional profiles of amino acids obtained by MEKC [17]. Passos and coworkers optimized the electrophoretic method paying special attention to the BGE composition (borate and sodium dodecylsulfonate –SDS– concentrations, pH and methanol content), the applied voltage and the running temperature. The best conditions in terms of peak resolution were obtained using a fused silica capillary (72 cm effective length and 50 μm ID), a 30 mM borate buffer at pH 10.1 containing 30 mM SDS and 5% (v/v) methanol as BGE. Temperature and voltage were set at 23°C and 25 kV. Detection was performed at 265 nm. The optimized method allowed the separation of 22 amino acids in pure standard mixtures. However, only six compounds (proline, glutamate, isoleucine, leucine, phenylalanine, and arginine) were used for the characterization of real juice samples due to strong matrix effects and loss of CE resolution. PCA models using amino acid concentrations allowed adulterated juices to be identified. In addition, it was concluded that glutamic acid seemed to be the most important variable to identify natural juices.

Navarro-Pascual-Ahuir et al. reported the optimization of two CZE methods for the determination of fumaric, isocitric, malic, tartaric and citric acids in fruit juices [19, 15]. In the first method, the separation was performed in a polyvinyl alcohol-coated capillary (64.5 cm, 56 cm effective length and 50 μm ID) with extended path light based on a "bubble cell" to enhance the method sensitivity. The BGE solution consisted of 200 mM phosphate buffer at pH 7.5, the temperature and the voltage were set at 25°C and - 15 kV, respectively. UV detection was performed at 200 ± 20 nm [19]. The second method used an uncoated fused-silica capillary (112.5 cm, 104 cm effective length and 75 μm ID) and a BGE containing 10 mM 2,6-pyridinedicarboxylic acid (PDC) as the chromophore and 0.5 mM cetyltrimethylammonium bromide (CTAB) to reverse electroosmotic flow direction. Separations were performed at - 25 kV and 10°C. In this case, indirect detection at 214 nm was used [19]. For both methods, intra- and inter-day precisions were lower than 5.3% and 1.5% for peak area ratio and migration time, respectively. In terms of detection limits, the preferred method depended on the analyte. However, as the authors stated, the indirect detection method did not require coated bubble cell capillaries thus resulting in an easier and more flexible and cost-effective option for routine determinations. In this case, LDA was applied to classify samples using concentration values of different organic acids in a variety of juices.

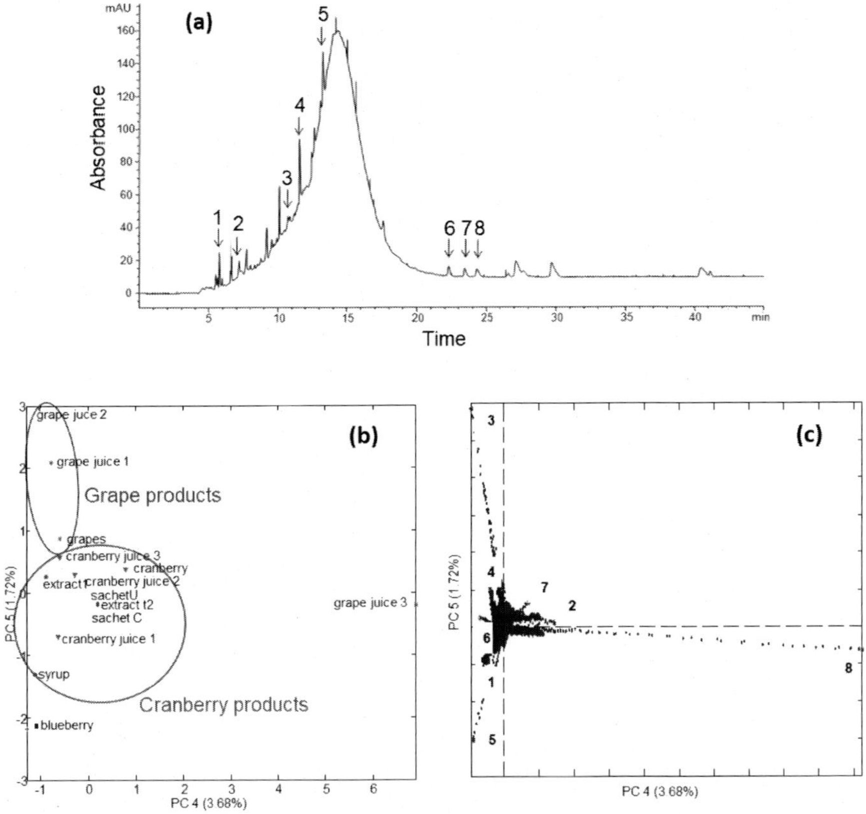

Figure 2. Characterization of cranberry and other fruits by PCA under a fingerprinting approach using the entire electropherograms. (a) Electropherogram recorded at 280 nm of a cranberry sample (numbers indicate some relevant features); (b) Scatter plot of scores of PC4 vs PC5; (c) Scatter plot of loading of PC4 vs PC5 (numbers correspond to peak features marked in the electropherogram).

As an alternative to classical gel electrophoretic techniques, CE has shown to be a suitable technique for the determination of proteolytic profiles of milk or cheese matrices. As the peptide content depends on the type of milk used for producing the dairy products and their maturation, the combination of CE with different chemometric data treatments was applied to determine the milk origin [16, 27, 28, 33] as well as to assess the cheese ripening time [31].

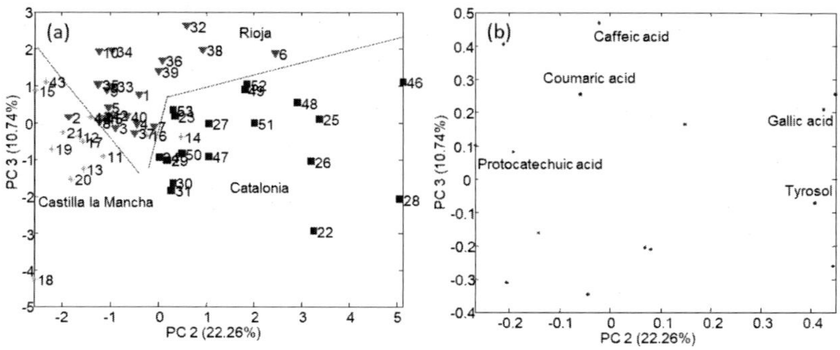

Figure 3. Characterization of Spanish red wines belonging to Catalonia, Rioja and Castilla la Mancha by PCA under a profiling approach using peak areas of targeted analytes and some unknown polyphenols. (a) Scatter plot of scores of PC2 vs PC3; (b) Scatter plot of loading of PC2 vs PC3 (non-labelled variables correspond to unknown compounds).

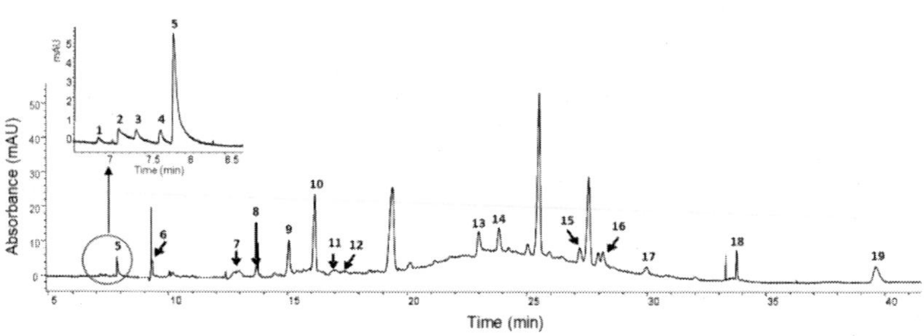

Figure 4. Electropherogram of a red wine sample recorded at 230 nm from in-capillary derivatization of biogenic amines with 1,2-nafthoquinone-4-sulfonate as the labelling reagent. Electropherogram contains peaks of polyphenols and amine derivatives. Peak assignation. 1 = agmatine; 2 = tryptamine (derivative); 3 = putrescine; 4 = cadaverine; 5 = tyramine (derivative); 6 = tyramine; 7 = resveratrol; 8 = tryptamine; 9 = phenylalanine + histamine; 10 = proline + alanine; 11 = catechin; 12 = glycine; 13 = syringic acid; 14 = ferulic acid; 15 = coumaric acid; 16 = vanillic acid; 17 = glutamic acid; 18 = 4-hydroxybenzoic acid; 19 = 2,4-dihydroxybenzoic acid.

For instance, Albillos and coworkers developed a CE method for protein analysis of the casein faction for predicting the ripening time of cheeses produced with different milk mixtures [31]. Two different CE modes were investigated in order to evaluate the proteolysis of the casein fraction. In the first method the separation was performed using a fused-silica capillary (57 cm, 50 cm to the detector window and 50 μm ID) and a 50 mM sodium phosphate with 6 M urea (pH 3.0), containing 0.05% (w/v) hydroxylpropyl methylcellulose (HPMC) buffer. The voltage applied was 18.5 kV. The second method used a neutral capillary (45 cm, 43 cm to the detector window and 50 μm ID) using a 0.32 M citric acid/0.02 M sodium citrate with 6 M urea (pH 3.0), containing 0.05% (w/v) HPMC as running buffer. Separation voltage was set at 25.09 kV. In both cases, UV detection at 214 nm was used. By visual inspection of the casein fraction electropherograms of samples at different ripening times, the authors were able to identify 21 and 16 peaks with the fused-silica or the neutral capillaries, respectively. An exploratory study by PCA allowed grouping samples according to their ripening time. Next, PLS and principal component regression (PCR) were used to determine cheese maturation using the areas of the identified peaks from samples obtained after 0, 16, 44, 73, 100, 133 and 167 days. Results showed a good correlation between ripening times predicted by cross-validation and real ripening times.

In another example, Rodriguez-Nogales and Vázquez developed a CE method to analyze the casein fraction for the determination of the bovine, ovine and caprine milk percentages in the manufacturing of Panela cheese [28]. The separation was performed in a fused-silica capillary (60 cm, 50 cm to the detector window and 50 μm ID) using a 50 mM phosphate buffer containing 0.05% HPMC and 6 M urea (pH 3.0) buffer. Separation voltage was set at 20 kV and detection was performed at 214 nm. In this case, 12 variables were chosen for chemometric analysis. The selection was based on the presence or not of the corresponding peak in the electropherogram obtained from pure or combined types of milk cheese samples. In order to assess the suitability of the method, authors used 60 cheeses from 10 different mixtures as training set and 18 cheeses from 3 different mixtures for validation purposes. PLS, PCR and multiple linear regression (MLR) were applied to the calibration samples using a full cross-validation process to select the optimal complexity of the models. Once the models were established, good prediction were obtained on the test samples with root mean square error of prediction (RMSEP) lower than 6.8, 7.7, 6.9 and 2.2 for PLS-1, PLS-2, PCR and MLR, respectively.

Franquet-Griell and coworkers developed a CZE method for the simultaneous determination of 20 polyphenols, including phenolic acids and flavonoids, in red wines of 3 Spanish regions, namely, Catalonia, La Rioja and Castilla La Mancha. Compounds were separated in a fused-silica capillary at a voltage of +25 kV using 30 mM sodium tretraborate (pH 9.2) with 5% isopropanol as the BGE solution. Matrix matched calibration was applied to quantify the targeted polyphenols in the wine samples. Compositional data consisting of peak areas of selected compounds was used for characterization and classification purposes. As shown in Figure 3, PCA results revealed the most significant analytes contributing to wine discrimination according to geographical origins, being tyrosol, gallic, protocatechuic, *p*-coumaric and caffeic acids some important descriptive compounds [25].

An interesting strategy to enrich the data set with the inclusion of new absorbing features was investigated by Garcia Villar et al. [29] proposing a CE method for the determination of biogenic amines (BAs) in wines. BAs occur in a wide variety of food products from the decomposition of protein components by the action of microorganisms. Apart from toxicological and organoleptic issues related to BAs, their compositional profiles have been utilized as potential descriptors of wine features and quality. Putrescine is the most abundant BA representing almost a 50% of the overall amine content, and histamine, tyramine and phenylethylamine are also common. Unfortunately, most of the BAs display low UV-vis absorption or poor native fluorescence properties so they need to be labelled to attain a sensitive detection. In this regard, the in-capillary derivatization of BAs with 1,2-naphthoquinone-4-sulfonate (NQS) demonstrated its great performance. As an example, Figure 4 shows that electropherograms contained complementary information from both polyphenolic and amine species that could improve the description about geographical origins and oenological practices. Hence, patterns dealing with the application of malolactic fermentation and wine ageing could be deduced more efficiently in this way.

CONCLUSION

This chapter introduces the great possibilities of the combination of CE with chemometrics in the field of food analysis to deal with characterization, classification and authentication studies. It has been emphasized that CE techniques are excellent to generate data of high quality and different complexity, basically consisting of compositional profiles (i.e., concentrations of targeted analytes) or electrophoretic fingerprints typically recorded by UV spectroscopy or mass spectrometry. Hence, the corresponding CE data results in a rich source of information which will provide chemical knowledge on food sample features such as geographical origin, product variety, elaboration process, etc. In general, CE has proved its performance for the determination of components in the range of concentrations from g L^{-1} to mg L^{-1} while its applicability to detect sub-trace components is more limited. Regarding the type of components under consideration, the approach has mainly been applied to the study of small organic molecules such as amines and amino acids, polyphenols and organic acids although its potentiality for the analysis oligomers and macromolecules should not be underestimated. By far, PCA is the most widely used chemometric method for tackling a preliminary sample exploration. Information gained from PCA is often the basis of further classification and quantification studies using more sophisticated mathematical tools such as PLS-DA, SIMCA, cluster analysis and PLS. More complex issues such as authentication and fraud detection can be tackled in this way. As a final comment, we reasonably expect that the number of applications of CE with chemometrics to food analysis will significantly rise in the next years.

REFERENCES

[1] Tzouros, N. E. & Arvanitoyannis, I. S. (2001). Agricultural produces: Synopsis of employed quality control methods for the authentication of foods and application of chemometrics for the classification of foods according to their variety or geographical origin. *Critical Reviews in Food Science and Nutrition* 4:287-319.

[2] Gad, H. A., El-Ahmady, S. H., Abou-Shoer, M. I. & Al-Azizi, M. M. (2013). Application of Chemometrics in Authentication of Herbal Medicines: A Review. *Phytochemical Analysis* 24:1-24.

[3] Mafra, I., Ferreira, I. M. P. L. V. O. & Oliveira, M. B. P. P. (2008). Food authentication by PCR-based methods. *European Food Research Technology* 227:649-665.

[4] Aparicio, R. & Aparicio-Ruiz, R. (2000). Authentication of vegetable oils by chromatographic techniques. *Journal of Chromatography A* 881:93-104.

[5] Di Stefano, V., Avellone, G., Bongiorno, D., Cunsolo, V., Muccilli, V., Sforza, S., Dossena, A., Drahos, L. & Vekey, K. (2012). Applications of liquid chromatography-mass spectrometry for food analysis". *Journal of Chromatography A*. 1259:74-85.

[6] Sentellas, S., Nunez, O. & Saurina, J. (2016). Recent Advances in the Determination of Biogenic Amines in Food Samples by (U)HPLC. *Journal of Agricultural and Food Chemistry* 64:7667-7678.

[7] Frazier, R. A. & Papadopoulou, A. (2003). Recent advances in the application of capillary electrophoresis for food analysis *Electrophoresis* 24:4095-4105.

[8] Kvasnicka, F. (2005). Capillary electrophoresis in food authenticity. *Journal of Separation Science* 28:813-825.

[9] Hernandez-Cassou, S. & Saurina, J. (2011). Derivatization strategies for the determination of biogenic amines in wines by chromatographic and electrophoretic techniques. *Journal of Chromatography B-Analytical Technologies in the Biomedical and Life Sciences* 879:1270-1281.

[10] Saurina, J. (2010). Characterization of wines using compositional profiles and chemometrics. *Trends in Analytical Chemistry* 29:234-245.

[11] Checa, A. & Saurina, J. (2013). Food Protected Designation of Origin: Methodologies and Applications. In: de la Guardia, M.; Gonzalvez, A. Book Series: *Wilson and Wilsons Comprehensive Analytical Chemistry* 60:123-143.

[12] Carranco, N., Farres-Cebrian, M., Saurina, J. & Nunez, O. (2018). Authentication and Quantitation of Fraud in Extra Virgin Olive Oils Based on HPLC-UV Fingerprinting and Multivariate Calibration. *Foods* 7:44.

[13] Pardo-Mates, N., Vera, A., Barbosa, S., Hidalgo-Serrano, M., Nunez, O., Saurina, J., Hernandez-Cassou, S. & Puignou, L. (2017). Characterization, classification and authentication of fruit-based extracts by means of HPLC-UV chromatographic fingerprints, polyphenolic profiles and chemometric methods. *Food Chemistry* 221:29-38.

[14] Erturk, A. S., Gode, F. & Bozdogan, A. E. (2017). Chemometric approach integrated with capillary electrophoresis for the investigation of public water system sources and quality. *International Journal of Environmental Science and Technology* 14:2459-2470.

[15] Navarro-Pascual-Ahuir, M., Lerma-Garcia, M. J., Simo-Alfonso, E. F. & Herrero-Martinez, J. M. (2017). Analysis of aliphatic organic acids in commercial fruit juices by capillary electrophoresis with indirect UV detection: application to differentiation of fruit juices. *Food Analytical Methods* 10:3991-4002.

[16] Mendes, T. D., Porto, B. L. S., Bell, M. J. V., Perrone, I. T. & de Oliveira, M. A. L. (2016). Capillary zone electrophoresis for fatty acids with chemometrics for the determination of milk adulteration by whey addition. *Food Chemistry*, 213:647-653.

[17] Passos, H. M., Cieslarova, Z. & Simionato, A. V. C. (2016). CE-UV for the characterization of passion fruit juices provenance by amino acids profile with the aid of chemometric tools. *Electrophoresis* 37:1923-1929.

[18] Pasquini, B., Orlandini, S., Goodarzi, M., Caprini, C., Gotti, R. & Furlanetto, S. (2016). Chiral cyclodextrin-modified micellar electrokinetic chromatography and chemometric techniques for green tea samples origin discrimination. *Talanta* 150:7-13.

[19] Navarro-Pascual-Ahuir, M., Lerma-Garcia, M. J., Simo-Alfonso, E. F. & Herrero-Martinez, J. M. (2015). Quality control of fruit juices by using organic acids determined by capillary zone electrophoresis with poly(vinyl alcohol)-coated bubble cell capillaries. *Food Chemistry* 188:596-603.

[20] Navarro-Pascual-Ahuir, M., Lerma-Garcia, M. J., Simo-Alfonso, E. F. & Herrero-Martinez, J. M. (2015). Rapid Differentiation of Commercial Juices and Blends by Using Sugar Profiles Obtained by Capillary Zone Electrophoresis with Indirect UV Detection. *Journal of Agricultural and Food Chemistry* 63:2639-2646.

[21] Mirasoli, M., Gotti, R., Di Fusco, M., Leoni, A., Colliva, C. & Roda, A. (2014). Electronic nose and chiral-capillary electrophoresis in evaluation of the quality changes in commercial green tea leaves during a long-term storage. *Talanta*, 129:32-38.

[22] Sanchez Arribas, A., Martinez-Fernandez, M., Moreno, M., Bermejo, E., Zapardiel, A. & Chicharro, M. (2014). Classification of Spanish white wines using their electrophoretic profiles obtained by capillary zone electrophoresis with amperometric detection. *Electrophoresis*, 35:1693-1700.

[23] Navarro, M., Nunez, O., Saurina, J., Hernandez-Cassou, S. & Puignou, L. (2014). Characterization of Fruit Products by Capillary Zone Electrophoresis and Liquid Chromatography Using the Compositional Profiles of Polyphenols: Application to Authentication of Natural Extracts. *Journal of Agricultural and Food Chemistry* 62:1038-1046.

[24] Hurtado-Fernandez, E., Contreras-Gutierrez, P. K., Cuadros-Rodriguez, L., Carrasco-Pancorbo, A. & Fernandez-Gutierrez, A. (2014). Merging a sensitive capillary electrophoresis-ultraviolet detection method with chemometric exploratory data analysis for the determination of phenolic acids and subsequent characterization of avocado fruit. *Food Chemistry*, 141:3492-3503.

[25] Franquet-Griell, H., Checa, A., Nunez, O., Saurina, J., Hernandez-Cassou, S. & Puignou, L. (2012). Determination of Polyphenols in Spanish Wines by Capillary Zone Electrophoresis. Application to Wine Characterization by Using Chemometrics. *Journal of Agricultural and Food Chemistry* 60:8340-8349.

[26] Montealegre, C., Garcia, M. C., del Rio, C., Marina, M. L. & Garcia-Ruiz, C. (2012). Separation of olive proteins by capillary gel electrophoresis. *Talanta* 97:420-424.

[27] Rodriguez-Nogales, J. M. & Vazquez, F. (2007). Application of electrophoretic and chemometric analysis to predict the bovine, ovine and caprine milk percentages in Panela cheese, an unripened cheese. *Food Control* 18:580-586.

[28] Rodriguez-Nogales, J. M. (2006). Approach to the quantification of milk mixtures by partial least-squares, principal component and multiple linear regression techniques. *Food Chemistry* 98:782-789.

[29] Garcia-Villar, N., Saurina, J. & Hernandez-Cassou, S. (2006). Capillary electrophoresis determination of biogenic amines by field-amplified sample stacking and in-capillary derivatization. *Electrophoresis*, 27:474-483.

[30] Scampicchio, M., Mannino, S., Zima, J. & Wang, J. (2005). Chemometrics on microchips: Towards the classification of wines. *Electroanalysis*, 17:1215-1221.

[31] Albillos, S. M., Busto, M. D., Perez-Mateos, M. & Ortega, N. (2006). Prediction of the ripening times of ewe's milk cheese by multivariate regression analysis of capillary electrophoresis casein fractions. *Journal of Agricultural and Food Chemistry* 54:8281-8287.

[32] Nunez, M., Pena, R. M., Herrero, C. & Garcia-Martin, S. (2000). Analysis of some metals in wine by means of capillary electrophoresis. Application to the differentiation of Ribeira Sacra Spanish red wines. *Analusis* 28:432-437.

[33] Vallejo-Cordoba, B. (1998). A chemometric approach to the detection of milk adulteration based on protein profiles determined by capillary electrophoresis. *Journal of Capillary Electrophoresis* 5:133-137.

[34] Vallejo-Cordoba, B. (1998). A chemometric approach to the detection of milk adulteration based on protein profiles determined by capillary electrophoresis. *Journal of Capillary Electrophoresis* 5:133-137.

In: Capillary Electrophoresis … ISBN: 978-1-53615-223-4
Editors: O. Núñez and X. Subirats

Chapter 3

CAPILLARY ELECTROPHORESIS IN FORENSIC ANALYSIS

Maykel Hernández-Mesa and Carmen Cruces-Blanco*
Department of Analytical Chemistry,
University of Granada, Granada, Spain

ABSTRACT

Forensic science is a strongly expanding research field and encompasses all the scientific disciplines that can be used to support law enforcement at any level. Just limiting the attention to the analytical aspect of forensic sciences, it involves an extremely wide application area that includes forensic toxicology, personal identification, analysis of explosives and gunshot residues, identification of fibers and hairs, inks and paints, etc. Consequently, the characterization and quantification of a wide range of compounds in a great variety of matrices must be addressed by forensic science. This fact supposes a big challenge faced by forensic scientists because the majority of forensic samples present a very complex nature and, moreover, their analysis requires very selective and sensitive analytical techniques.

* Corresponding Author's Email: mcruces@ugr.es.

In the last decades, capillary electrophersis has shown to be a powerful tool that provides solutions to almost any analytical question arising in forensic laboratories. As result, it has become an complementary and essential separation technique in this field, particularly for the analysis of forensic DNA. To illustrate this fact, this chapter offers a comprehensive overview of the most recent applications of CE in forensic analytical cemistry, covering its different separation modes and its hyphenation with different detection systems.

Keywords: forensic analysis, capillary electrophoresis, drugs of abuse, forensic DNA, explosives, gunshot residues, dyes

INTRODUCTION

Forensic science comprises a wide range of scientific areas that, simultaneously, have to satisfy scientific, juridical and practical demands. It involves a broad variety of disciplines such as pathology, entomology, odontology, etc., and also includes chemistry. Forensic analytical chemistry is the application of analytical chemistry for law enforcement and supports forensic investigations by providing scientific evidence. Under this context, it covers the qualitative and/or quantitative analysis of forensic samples collected at the crime scene, from a suspect and/or from a victim in the course of a criminal investigation.

Analytical instruments, especially those related to separation techniques, are the principal tools of forensic chemistry. Capillary electrophoresis (CE) is one of the most rapidly growing analytical technologies in the last three decades. Since the launch of the first commercial CE instrument in 1988, numerous books and reviews have been published about this technique [1, 2, 3]. It covers a family of related methodologies in which separations are carried out in narrow bore capillaries under the influence of an electric field. CE is an attractive technique for forensic laboratories due to its exceptional power of separation and resolution, short analysis time, low solvent consumption, and minimum sample requirements [4, 5].

Despite its advantages, the implementation of CE has been limited by its poor concentration sensitivity, particularly when applied with on-column ultraviolet (UV) detection, due to the low sample injection volume and short optical path-length. In addition, low selectivity is also attributed to UV detection. Nevertheless, UV detection is usually employed in CE-based methods because it is standard on commercial instruments. In this sense, they can also be fitted with ultraviolet diode array detection (UV-DAD) for the simultaneous acquisition of spectra at different wavelengths, improving method selectivity [6]. On the other hand, the lack of sensitivity can be overcome using detection modes more sensitive than UV detection, such as fluorescence detection. However, on-column fluorescence detection can also present sensitivity limitations arising from the short path-length provided by CE capillaries and from the weak fluorescence signals produced by many analytes [7]. In such cases or in those where analytes do not exhibit native fluorescence, reactions of derivatization can be carried out using the proper derivatization agents (e.g., fluorescein-5-isothiocyanate (FITC)) in order to enhance fluorescence. Since few analytes are naturally fluorescent, pre-column derivatization is often applied. Laser induced fluorescence (LIF) detection is more sensitive but is also more complex from an instrumental point of view and requires the appropriate laser source for the detection of an analyte or a group of analytes. Additionally, electrochemical detection, in particular amperometric and conductimetric detection, has also been applied in combination with CE, leading to encouraging results [8]. However, this strategy has not been widely employed for forensic purposes.

In the last years, the hyphenation of CE with mass spectrometry (MS) has emerged as a powerful methodology for the analysis of forensic samples because it combines the high speed and efficiency of CE with the universality, selectivity and sensitivity inherent to MS [9]. It is clear that CE-MS is already a mature technique but, to reach this point, a lot of research has been required in order to solve several difficulties, especially those related to CE-MS interface. In must be taken into account that CE-MS interface must ionize analytes coming from the aqueous background electrolyte (BGE) solution into a gas phase and, the most important issue, it must guarantee that the electric circuit of the electrophoretic system is closed. Although different ionization interfaces have been proposed for CE-MS coupling, electrospray ionization (ESI) interface remains as the most commonly used in CE-MS methods [10]. It offers high ionization efficiency for a wide range of compounds and is a soft ionization, so the fragmentation of analytes during ionization is minimized.

The versatility of CE as analytical technique is not only a consequence of multiple detection modes that can be employed, but also because the separation of compounds can be achieved by a wide range of electrophoretic approaches based on different separation mechanisms. In general, capillary zone electrophoresis (CZE) is the most frequently used CE separation mode and in which charged analytes are separated based on their charge-to-size ratio. In CZE, the pH of BGE solution is a key factor in the selectivity. In this sense, a separation medium with high buffer capacity is required in order to achieve reproducible separations. If BGE solution is prepared in an organic solvent (i.e., acetonitrile (MeCN), methanol (MeOH), ethanol (EtOH), etc.) instead of water, separations take place under non-aqueous capillary electrophoresis (NACE) mode. On the other hand, micellar electrokinetic chromatography (MEKC) combines the principles of electrophoresis and chromatography allowing the separation of both neutral and charged solutes by the addition of surfactants at concentration levels higher than their critical micellar concentration (CMC) to the BGE solution. In addition, capillary electrochromatography (CEC) combines the high efficiency of CE and the high selectivity of liquid chromatography (LC). In CEC, charged and neutral analytes are separated by their distribution between a mobile and a stationary phase while the mobile phase (i.e., BGE solution) is propelled through the capillary by the electroosmotic flow (EOF). Packed, monolithic and open-tubular capillaries can be used as separation column in CEC. Furthermore, charged molecules can also be separated based on sieving mechanisms using capillaries filled with porous gel matrix. This CE mode is known as capillary gel electrophoresis (CGE).

Based on the foregoing, CE technology has great potential to be applied in those cases where an analytical separation technique is needed. In this sense, several CE-based applications have been developed in the field of forensic analysis. It is a complementary technique to traditional LC and MS methodologies, and offers interesting performance characteristics, especially when is combined with MS. This chapter provides a comprehensive overview of the CE methods focused on forensic analysis that have been recently developed. It covers the most relevant applications in the field, including the analysis of drugs of abuse, explosive compounds and gunshot residues (GSRs), forensic DNA and dyes in inks, fibers and artwork. With the literature found in the early twenty-first century, the potential of CE to become an indispensable tool for forensic scientists is well demonstrated.

APPLICATIONS OF CAPILLARY ELECTROPHORESIS IN FORENSIC ANALYSIS

Forensic Toxicology

Nowadays, forensic scientists have a great responsibility since their evidence in court can have a crucial bearing on the conviction or otherwise of a defendant [11]. Under this context, the analysis of drugs remains a major activity in forensic laboratories. Legislation prescribes a variety of charges and penalties for different illegal substances in relation with their concentration levels. Therefore, it is really important to correctly identify and quantify the possible presence of these types of substances in forensic samples [12]. Toxicological analysis is mainly aimed at the search of selected substances that are part of a scheduled list of toxicants or harmful compounds where illicit drugs are included. Forensic toxicology comprises the scientific study of these substances in relation to the law and the work of forensic toxicologists is to detect and quantify them in very complex matrices such as blood, plasma, urine, tissue, saliva, vitreous humor and hair. In this sense, the identification and determination of a broad range of drugs of abuse and their metabolites in all these different biological matrices has become a challenging task for analytical chemists.

From a historical point of view, the use of drugs of abuse began after the Second World War, particularly during the 1960s, when cannabis became a symbol of the hippie revolution. In the 1980s, the consumption of two well-known drugs, namely cocaine (COC) and crack, reached a peak. Ten years later, the consumption of other recreational drugs (e.g., 3,4-methylenedioxymethamphetamine (MDMA), commonly known as ecstasy) grew rapidly. In the last twenty years, new drugs have appeared in the illicit market. These molecules are commonly referred to as 'new psychoactive substances' and include synthetic cannabinoids, cathinones and tryptamine analogs of psilocin [13]. The entry of these new psychoactive substances into the recreational drug market is causing a serious health problem worldwide that requires the adoption of new legal measures for their control.

Drugs are classified in acidic, neutral or basic compounds. Basic drugs of abuse are the largest type of drugs and their analysis is the major activity of forensic laboratories. They are basic amine compounds and include narcotics such as morphine, stimulants such as COC and amphetamines, and tranquillizers and antidepressants such as chlorpromazine and amitryptilene.

The analysis of illicit drugs in biological samples usually requires two steps: preliminary screening or presumptive testing and confirmatory testing. Preliminary screening can be carried out in situ with, for example, spot test kits, but these analyses are less precise and only indicate that an illicit drug or type of substances may be present in the sample. Therefore, it is often necessary to send the samples to a forensic laboratory for confirmatory testing.

Confirmatory testing requires the use of instrumental analysis, usually gas chromatography (GC) or high performance liquid chromatography (HPLC) coupled with MS. Beside these techniques, which are widely used in private and governmental laboratories, electromigrative separation techniques such as CE, have gained recognition as separation techniques in forensic science in the last two decades [14, 15]. Since the potential of CE for forensic analysis was first demonstrated in the early 1990s, it has been applied to a wide range of analytes of forensic interest including illicit drugs in biological and non-biological samples [16, 17, 18, 19, 20].

In many instances, the choice of CE as an alternative technique to chromatographic methods is justified by the minimal amount of sample required. Additionally, owing to the different CE modes (i.e., CZE, MEKC, CEC, capillary isoelectric focusing (CIEF), capillary isotachophoresis (CITP), CGE) and the detection systems available (i.e., UV spectroscopy, fluorescence, luminescence, MS, and electrochemical methods in less extension), the applicability of CE have been extended to a wide range of chemical substances. This is the main reason for which some laboratories have adopted CE as a complementary technique to routine chromatographic methods, and also due to its high efficiency and low solvent consumption [21, 22].

Despite the high separation potential of CE, sensitivity still remains as the major drawback for its application to illicit drug analysis. This problem is mainly due to the short optical path-length of the most frequently employed detector in CE methods (i.e., UV-Visible detector) and the small sample volume injected into the capillary. In order to overcome this problem, several on-line pre-concentration techniques based on different stacking mechanisms are currently exploited in many application methods [23]. Pre-concentration can also be achieved through sample preparation, which on the other hand, is usually required to carry out for sample clean-up due to the high complex nature of biological matrices of forensic interest.

Liquid-liquid extraction (LLE) is the most commonly used sample treatment in forensic toxicology laboratories for the extraction and pre-concentration of analytes from toxicological matrices. Ten years ago, Rezaee et al. [24] introduced a miniaturized LLE technique, the so-called dispersive liquid-liquid microextraction (DLLME), which has been widely used for the extraction of several forensic drugs and poisons such as amphetamines, cannabinoids, antidepressants, hallucinogens, etc. from numerous biological samples [25]. Furthermore, solid phase extraction (SPE) is another long-used technique and is characterized by high recovery efficiency, a moderate solvent consumption and satisfactory reproducibility [26]. In recent years, solid phase microextraction (SPME) has emerged as an alternative to SPE, showing its potential in forensic applications [27]. SPME is a solvent-free procedure in which analytes are extracted from a gaseous or liquid sample by absorption in, or adsorption on, a fiber coated with a polymeric stationary phase.

Under this context, blood and urine samples frequently serve as sources of biological specimens for drug analysis. The preparation and handling of these samples take advantage from the experience of the clinical area, being a well-established starting point on which the forensic analyst must define the specific features of each case. From a forensic perspective, the uniformity is a key advantage of the analysis of blood samples, either as serum or plasma, because they are largely unaffected by confounding factors such as age, sex, diet, diurnal cycles and stress. On the other hand, urine is the most commonly used sample in forensic analysis because, unlike blood, sample collection is non-invasive, is easy to be performed under conditions that prevent adulteration and substitution, and provides direct evidence for the short-term use of illicit drugs. However, targeted analytes such as methamphetamine remain in urine no longer than 10 days after use, which is why urine analysis cannot distinguish between habitual or first-time users [28].

Nowadays, besides blood and urine, hair analysis is gaining increasing popularity in forensic toxicology as a tool for investigating past and chronic exposure to illicit drugs [29, 30]. Like urine, hair belongs to the realm of non-invasive samples. The major advantages of hair testing are the ease of its collection and the large detection window to identify the consumption of drugs of abuse. The presence of these substances in hair samples can be detected from 3 days to years after administration, depending on the length of the hair shaft. Generally, scalp hair grows at an average of 1 cm per month and this means that each centimeter of hair records about 1 month of the individual´s use or exposure to a drug [31]. On the contrary, the detection window for the majority of drugs in urine and blood samples ranges from hours to 2–4 days. Furthermore, drugs as well as their metabolites can be determined in hair samples [30, 32]. In this regard, CE offers clear advantages over current chromatographic techniques due to the minimal amount of sample required for the analysis, which is a crucial point for hair samples. Tagliaro et al. proposed the first CE-method applied to the determination of drugs of abuse in hair [33], and their group was responsible of the acceptance of hair analysis in a panel of clinical and laboratory tests [34].

The aim of this section is to report the application of CE in forensic toxicology for the analysis of drugs of abuse in biological specimens. In the same way as chromatography techniques, CE is suitable to be coupled with different detection techniques, and MS and UV-detection are the most frequently used. This section is organized based on the different detection systems that are more frequently used in forensic laboratories when CE is applied as separation technique.

Capillary Electrophoresis with UV Detection

CE with UV-absorbance detection is the methodology most widely used for forensic toxicological analysis of drugs of abuse because it is standard on commercial instruments, which can also be fitted with a DAD detector for the simultaneous acquisition of spectra [6]. However, due to the lack of sensitivity of this detection mode for the quantification of illicit drugs in biological samples, CE-UV methods usually involve the application of on-line and/or off-line pre-concentration methodologies, as discussed above.

In the field of toxicological analysis, one of the first applications of CE-UV was developed by Werny and Thormann in 1991 for the analysis of different drugs of abuse, including opioids, COC metabolites, amphetamines and hypnotics, in urine samples [35]. In this work, samples (5 mL) were submitted to SPE, analyzed by MEKC and detected at 195 nm. As a consequence, drug concentrations down to the ng/mL level were detected in this biological matrix. In general, off-line SPE is generally used to achieve high sample concentration and reach low limits of detection (LODs), and a wide range of SPE-based methods focused on the determination of drugs of abuse can be found in the literature. For example, a SPE-CZE-UV method using Oasis HLB cartridges (Waters, Milford, MA, USA), which consist of a reverse phase sorbent with hydrophilic/lipophilic balance and commonly used in SPE procedures, has been proposed for the analysis of two opium alkaloids (noscapine and papaverine) and two anesthetics (lidocaine and bupivacaine) in spiked urine samples [36]. The developed sample treatment was very specific for these compounds, obtaining recoveries between 81-

107%. Moreover, in order to avoid the lack of sensitivity of UV detection, Alnajjar et al. have developed a sequential injection-solid-phase extraction (SI-SPE) system for the analysis of nineteen different drugs of abuse in urine samples [37]. Using this device, matrix clean-up, extraction, and pre-concentration of analytes were performed onto an octadecyl (C18) cartridge. Subsequently, the separation of compounds was performed by CE-UV using a BGE solution consisted of 100 mM phosphate (pH 6), 20 mM β-cyclodextrin (β-CD), 5% (v/v) MeCN and 20% (v/v) isopropanol. Finally, analytes were detected at 214 nm, achieving LODs in the range of 5-30 ng/mL. In-line SPE has also been used as an enrichment technique in combination with CE-DAD equipment for the pre-concentration and separation of four drugs of abuse (i.e., 2-ethylidene-1,5-dimethyl-3,3-diphenylpyrrolidine (EDDP), codeine (COD), 6-acetyl-morphine (6-AM), and COC) in water samples, which were directly analyzed without any off-line sample treatment [38]. Relative recoveries between 85 and 97% were obtained and LODs ranged between 50 and 200 ng/L.

Due to the complexity of biological samples and the sensitivity required in forensic analytical methods, it is very important to carry out sample clean-up and pre-concentrate trace amounts of analytes using an efficient extraction method before sample separation and detection. As alternative to SPE-based methods, DLLME has been widely applied to the extraction of drugs of abuse prior to analysis by CE methodologies using UV detection [39].

Regarding the analysis of urine samples in forensic toxicology, the concentration of creatinine is normally used as a standardization tool for the quantitative measurement of therapeutic, illicit drugs and/or xenobiotics. Recently, it has been proposed a robust and reliable CE method using direct UV detection at 200 nm that meets the strict requirements of forensic analysis and is very useful to test the possible adulteration or dilution of urine samples under toxicological screening [40].

As can be deduced from the literature, the most frequently biological sample analyzed by CE with UV detection is urine. In this sense, CE-UV methods intended for urine analysis usually involve the application of an on-line sample pre-concentration procedure to enhance the sensitivity required in this type of analysis. Field-amplified sample injection (FASI) is a classical on-line pre-concentration method that occurs when samples are electrokinetically injected from an injection medium less conductive than the BGE solution (i.e., normally diluted BGE solution). It allows the injection of a great amount of sample without any loss of peak resolution [41]. A novel DLLME-FASI-CZE-UV method has been recently developed for the analysis of three drugs of abuse (i.e., lysergic acid diethylamine (LSD), phencyclidine (PCP), and MDMA) in urine samples [42]. DLLME was employed as sample treatment, and the main parameters affecting DLLME were optimized by an experimental design (i.e., central composite design) using the response surface methodology (RSM). As a consequencc of the sample pre-concentration achieved by FASI, LODs of 1.00, 4.50, and 4.40 ng/mL were obtained for MDMA, PCP and LSD, respectively. Figure 1 shows the electropherograms resulted from the analysis of urine samples fortified with the three drugs before and after DLLME, and a blank sample of urine treated by DLLME under the optimized conditions.

In addition to FASI, sweeping is also an on-line pre-concentration strategy widely used in MEKC separations. It was introduced in 1998 by Quirino and Terabe, and is based on a focusing effect that occurs when a sample is hydrodynamically injected in a solution devoid of micelles [43]. During the separation, the pseudostationary phase of BGE solution penetrates the sample zone, 'sweeps' and accumulates the analytes. In addition, the combination of sweeping with other pre-concentration approaches can also provide an additive effect on sensitivity. For example, several drugs of abuse (i.e., COD, M, K, MA) have been simultaneously analyzed by cation-selective exhaustive injection (CSEI) in combination with sweeping and MEKC in real human urine samples at detection limits down to ng/mL levels, and without requiring any sample treatment [44]. In comparison with CZE, the concentration sensitivity achieved by the proposed CSEI-sweeping-MEKC method and related to the analysis of MA could be increased up to 6000-fold.

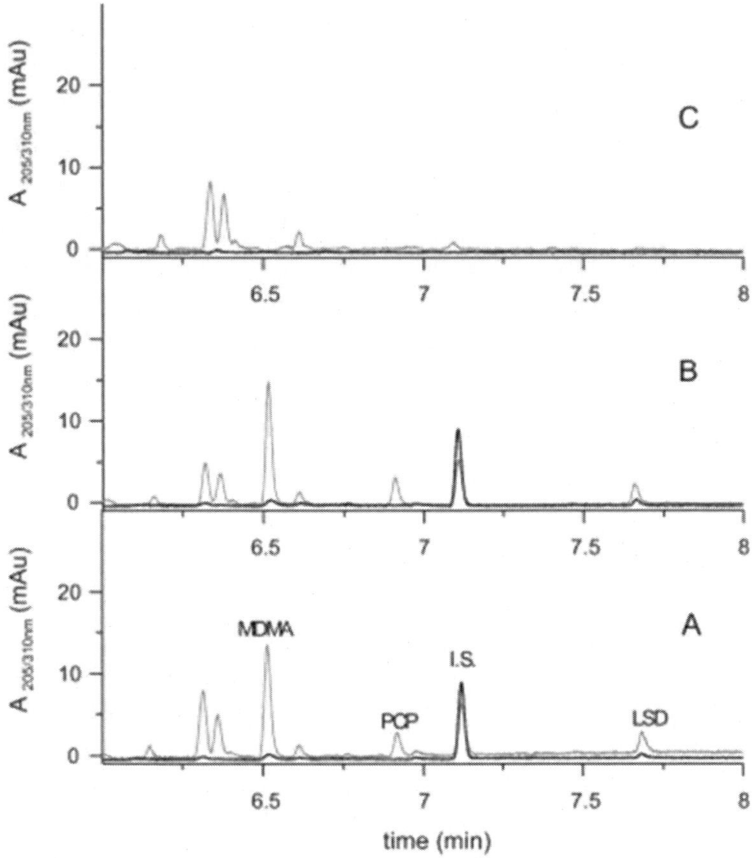

Figure 1. Electropherograms at 205 (grey) and 310 nm (black) for urine samples submitted to the DLLME treatment and spiked with 47.5 ng/mL of MDMA, PCP and LSD (A) directly in the sample solution and before applying the treatment, (B) just before injection and (C) blank sample. Reproduced with permission from [42].

Cannabis is one of the most commonly drug of abuse in the world. Its main psychoactive component is Δ(9)-tetrahydrocannabinol (THC) and, in humans, it is metabolized primarily into 11-hydroxy-9-tetrahydrocannabinol (THC-OH) and further to 11-nor-9-carboxy-9-tetrahydrocannabinol (THC-COOH). Recently, a large volume sample injection (LVSI)-anion selective exhaustive injection (ASEI)-sweeping-MEKC method has been developed for the sensitive and rapid analysis of THC and its major metabolites [45]. In comparison to conventional MEKC, concentration sensitivity was enhanced up to 2000-fold and analysis time was drastically reduced in comparison to the complete analysis time (from 4 to 5 hours) required by chromatographic techniques.

Benzodiazepines (BZDs), which constitute a group of drugs of abuse widely consumed, are frequently combined with the administration of COC, MA or MOR by many addicts. This fact makes difficult to detect them within a reasonable short time due to their similar structures. Moreover, due to the neutral properties of BZDs at most pH values, it is not possible to use electrokinetic injection for sensitivity enhancement when CE methodologies are applied to their analysis. For this reason, hydrodynamic sample injection in combination with stacking with reverse migration micelles (SRMM) has been evaluated for the determination of BZDs by MEKC-UV, in presence of other drugs of abuse and in urine samples from addicts [46]. The use of a chemometric experimental design allowed a systematic optimization process of the separation conditions. Additionally, a large volume sample stacking (LVSS)-sweeping-MEKC method has also been developed for the pre-concentration and analysis of ten drugs of abuse, including BZPs, in urine samples [47]. Sensitivity was increased more than 200-fold by the proposed stacking method. This method was designed to serve as a fast tool for screening abused drugs.

In criminal investigations, it is necessary to discriminate illegal abuse from legitimate clinical use of amphetamines. Despite there are many methods for analysis of amphetamines and their metabolites, not all the methods have the possibility of carrying out chiral analysis. Under this context, CE is a powerful and low-cost tool for the chiral analysis of amphetamines. Iwamuro et al. have evaluated the applicability of a poly(vinyl alcohol) (PVA)-coated capillary for the chiral analysis of methamphetamine (MA) and related compounds in urine samples [48]. In addition, chiral selectors such as cyclodextrins (CDs) can be added to the BGE solution in order to achieve enantiomeric selectivity. The use of CDs improves separation efficiency due to their hydrophobic cavity. The selectivity can be easily tuned by adjusting types and concentrations of chiral selectors. CDs have been employed for the analysis of heroin metabolites in human urine using CZE-UV [49], being a sensitive and reproducible screening method for forensic analysis. In addition, this method used a SPE procedure as sample treatment that provided clean extracts and resulted in high recoveries for all heroin metabolites.

Furthermore, CEC has also been applied to the forensic analysis of urine samples. An open tubular-CEC method using single-drop microextraction (SDME) as sample treatment has been developed for the analysis of seven drugs in urine of five different horses collected before a horse race [50]. SDME allowed to reach an excellent clean-up of the horse urine samples and recoveries between 80 and 108%.

In recent years, the analysis of non-conventional biological samples for forensic purposes has grown, replacing or complementing traditional methods focused on the analysis of urine and blood. In this sense, each biological matrix has its own peculiarities and present advantages and disadvantages of use. One of these non-conventional matrices is vitreous humor that constitutes a relatively simple sample in analytical terms compared to blood and urine. Costa et al. have developed the first CE-DAD method for the analysis of twelve drugs of abuse (i.e., 6-AM, AM, MA, 3,4-methylenedioxymethamphetamine (MDMA), 3,4-methylenedioxyamphetamine (MDA), 3,4-methylenedioxy-ethylamphetamine (MDEA), ketamine, COC, cocaethylene, lidocaine, morphine (MOR) and heroin) in human vitreous humor [51]. Eighty-seven samples were analyzed by CE-DAD and the results were confirmed by LC-tandem mass spectrometry (MS/MS). COC was detected in 7.1%, cocaethylene in 3.6%, lidocaine in 2.4% and ketamine in 1.2% of the total number of analyzed samples.

Among non-conventional forensic samples, hair samples offer particular advantages. CE began to be applied to the analysis of drugs of abuse in hair after the work of Tagliaro et al. [33], in which the determination of COC and MOR in hair by CZE-UV was proposed. Efficient separation as well as accurate and precise determinations (intra-day relative standard deviations (RSDs) in the range of 3 to 5%) was achieved for both drugs in hair extracts, avoiding the interferences of more than 90 therapeutic drugs and drugs of abuse. Baciu et al. have recently reported the results of a study carried out in four patients undergoing a racemic methadone (MTD) treatment program [52]. The CE-UV method was suitable for the simultaneous determination of 6-AM, benzoylecgonine (BZE), COC, COD, MOR and the enantiomers of MTD in human hair. Samples were firstly submitted to pressurized liquid extraction (PLE) and sample clean-up was accomplished by in-line SPE. In addition, α-CDs were added to the BGE solution for achieving the separation of enantiomers. Recoveries greater than 80% were obtained.

Despite the many advantages offered by the forensic analysis of hair samples, a high sensitive analytical method is essential for the determination of drugs in hair due to their low concentration in this type of matrix as well as the low availability of this type of sample in some cases. Consequently, several on-line pre-concentration methods have been evaluated for enhancing the concentration sensitivity achieved by CE-based methods focused on hair analysis. For example, a cation-selective exhaustive injection (CSEI)-sweeping-MEKC-UV method has been developed for the simultaneous determination of MA, ketamine, MOR and COC in real hair samples taken from addicts [53].

Many drugs, alone or in combination, have the potential to be used to facilitate sexual assault. They are known as drugs of abuse of sexual attack (DFSA) and includes alcohol, BZDs and, specially, liquid ecstasy (γ-hydroxybutyric acid, GHB). The type of forensic samples that are appropriate for toxicological examination depends on the time since the sexual attack. Early after the assault, urine and blood can be analyzed but the analysis of these drugs in hair, alcoholic and non-alcoholic beverages, is recommended after a longer period of time. A rapid CZE method with DAD detection was developed for the simultaneous determination of nine BZDs in spiked beverages in 6.5 min [54]. Different alcoholic and non-alcoholic beverages were spiked with each drug at a concentration designed to simulate a prescription tablet but no sample pretreatment was required prior analysis. MEKC has also been used for the separation of several DFSA and club drugs using an anionic surfactant (i.e., sodium dodecyl sulfate, SDS) as pseudostationary phase [55]. All the studied analytes, including BZDs and GHB, were baseline resolved using a BGE solution consisted of 30 mM SDS in borate buffer (pH 9.2) with 10% (v/v) MeCN. The determination of these substances in beverages samples only required a one-step LLE with ethyl acetate as sample treatment.

Table 1 shows the most relevant CE-UV methods developed for the determination of drugs of abuse in different forensic matrices.

Table 1. Applications of CE-UV to the analysis of drugs of abuse

Analyte	Analytical technique	LODs	Type of sample	Reference
COC and MOR	CZE-UV	0.15 ng/mg	Hair	[33]
BZE, MOR, heroin, MA, COC, AM, COD, MTD, flunitrazepam, oxazepam, diazepam, methaqualone	SPE-MEKC-UV	100 ng/mL	Urine	[35]
AM, MA, ephedrine, psilocin, MTD, PCP, COC, cocaethylene, pheniramine, 6-AM, COD, MOR, diphenhydramine, oxycodone, noscapine, fentanyl, heroine, thebaine, papaverine	SPE-CE-UV	5 - 30 ng/mL	Urine	[37]
EDDP, COC, COD, 6-AM	SPE-CE-UV	70 - 270 ng/L	Tap and river water	[38]
Lidocaine, bupivacaine, noscapine, papaverine	SPE-CZE-UV	0.06 - 0.12 μg/mL	Urine	[36]
Flunitrazepam, alprazolam, diazepam, nitrazepam, 7-aminoflunitrazepam, temezepam, 7-aminonitrazepam, oxazepam, 7-aminoclonazepam	CZE-DAD	2.7 - 41.5 μg/mL	Beverages	[54]
GHB, γ -butyrolactone, flunitrazepam, alprazolam, traizolam, midazolam, chlordiazeposide, lorazepam, prazepam, clonazepam, sulfanilic acid (internal standard, i.s.)	MEKC-UV	0.979 – 152 μg/mL	Beverages	[55]
MA, MOR, COD, alprazolam, clonazepam, oxazepam, diazepam, flunitrazepam, nitrazepam, ketamine	LVSS-sweeping-CE-UV	7.5 - 30 ng/mL	Urine	[47]
MOR, MA, COD, ketamine, alprazolam, clonazepam, diazepam, flunitrazepam, α-hydroxyalprazolam, nordiazepam, 7-aminoflunitrazepam, 7-aminoclonazepam, oxazepam, N-demethylflunitrazepam	SRMM-MEKC-UV	20 - 50 ng/mL	Urine	[46]

Analyte	Analytical technique	LODs	Type of sample	Reference
MA, MOR, COD, ketamine	CSEI-sweeping-MEKC-UV	5 - 15 ng/mL	Urine	[44]
THC, THC-OH, THC-COOH	LVSI-ASEI-sweeping-MEKC-UV	0.5 - 10 ng/mL	Urine	[45]
Ephedrine, ketamine	DLLME-CE-UV	0.21 - 0.39 µg/mL	Urine	[39]
MDMA, PCP, LSD	DLLME-FASI-CZE-DAD	1.0 - 4.5 ng/mL	Urine	[42]
MA, AM, MDMA, MDA, MDEA, norephedrine, methylephedrine, ephedrine, dimethylamphetamine, *p*-hydroxymethamphetamine	CE-UV	3 - 5 ng/mL	Urine	[48]
Ephedrine, COC, strychnine, MOR, caffeine, theophylline, piroxicam	Open tubular-CEC-UV	0.94 - 17.64 ng/mL	Horse urine	[50]
MOR, COD, 6-AM, normophine	CZE-UV	30 - 40 ng/mL	Urine	[49]
AM, MA, MDA, 6-AM, MDMA, MDEA, ketamine, COC, MOR, heroin, cocaethylene, lidocaine	CE-DAD	1 - 5 ng/mL	Vitreous humor	[51]
MA, ketamine, MOR, COD	CSEI-sweeping-MEKC-UV	50 - 200 pg/mg	Hair	[53]
COC, MOR, COD, 6-AM, MTD, BZE	In-line-SPE-CE-UV	0.10 - 1.00 ng/mg	Hair	[52]

Capillary Electrophoresis with MS Detection

As discussed previously, UV-Visible detection has by far been the most widely used in CE methods, but its sensitivity is always relatively low for the requirements of forensic toxicologists. Although fluorescence detection is more sensitive, a complex derivatization step prior to detection is usually required, since few drugs of abuse are naturally fluorescent. Moreover, the abovementioned detection systems provide poor structure information of the analytes under study, which is, in many circumstances, very necessary. On the contrary, MS provides very useful information about the analytes of interest that can be used for their unequivocal identification. Consequently, high efforts are made to develop CE-MS methods for the analysis of drugs of abuse.

Within this framework, MS provides precise information about chemical formulas, but isomeric compounds cannot be distinguished based on their mass-to-charge (*m/z*) ratio. Therefore, chiral separation must be carried out prior to MS detection. CE using chiral selectors, mainly CDs, and coupled with MS is a good strategy to distinguish between congeners of drugs of abuse. MA has an asymmetric carbon on its structure, so there are two enantiomers. Dimethylamphetamine (DMA) is metabolized to MA, AM and *p*-hydroxymethamphetamine and these metabolites are excreted into the urine. Consequently, chiral analysis of urine samples is necessary for the proof of MA use [56].

In general, CE-MS methods for forensic analysis can follow the same or similar strategies than those used by CE-UV methods. For example, in-line SPE-CE-MS has been used for the analysis of EDDP, COD, hydrocodeine and 6-AM [57]. The proposed method was applied to the study of urine samples diluted 1:1 (v/v) and spiked with these four drugs in the linear range from 0.08 to 10 ng/mL. This method enabled the simple and effective determination of these drugs of abuse in urine samples at the levels encountered in toxicology and doping laboratories.

Normally, ESI is used for CE-MS hyphenation. CE coupled via ESI to MS, with a very simple sample preparation, has shown to be very efficient for the screening and quantitation of COC and five metabolites in human urine [58]. The proposed method presented good linearity (dynamic range from 250 to 5000 ng/mL) and intra-day precision (RSDs $< 10\%$). In general, CE-ESI-time-of-flight (ToF)-MS is widely used for screening purposes, due to the high mass resolution and accuracy of ToF mass spectrometers that allow the identification of unknown compounds. For example, a fast and sensitive CE-MS two-step workflow was developed for the screening and quantitation of opiates, amphetamines and derivatives, and COC and metabolites, in urine samples [59]. Firstly, samples were submitted to screening by CE-ESI-MS/MS. The quantitation of compounds was further performed by CE-ESI-MS/MS with a triple quadrupole (QqQ) analyzer working in selected reaction monitoring (SRM) mode. COC and MTD detection in urine samples was determined to be accurate over the range of 10–1000 ng/mL and 21–1000 ng/mL, respectively. Another CE-ToF-MS method using an ESI source and a coaxial sheath-flow interface in combination with DLLME has been employed for the screening of a wide range of toxicological compounds in urine samples [60]. With a pre-concentration factor of more than 130, this highly sensitive DLLME-CE-ESI-ToF-MS method allowed the detection of 30 toxicological compounds (i.e., amphetamines and their derivatives, opiates, COC and its metabolites and pharmaceutical) in urine samples with LODs at sub-ng/mL levels.

Cannabis is one of the most commonly consumed illicit drugs worldwide. The increasing number of arrests due to the extended use of cannabis has resulted in an increased demand for forensic analysis of this drug of abuse. THC is the psychoactive substances and is mainly metabolized into THC-COOH. In a second phase, it is metabolized to THC-COOH-glucuronide that is excreted into the urine as a major metabolic end product of THC. For such a reason, different analytical methods have been developed for their simultaneous determination. Iwamuro et al. have proposed a simple and rapid method for the direct analysis of the two major metabolites of THC in urine samples by CE-MS [61]. Sample treatment consisted of sample dilution with MeOH. Good results were obtained in terms of precision for migration times and peak areas (RSDs < 0.36 and 25%, respectively).

As it was indicated before, a very serious health and legal problem is currently caused by the increasing commercialization of the so-called new psychoactive substances, which may often not be detected by standard screening methods. Among the wide group of new psychoactive substances, synthetic cathinones are one of the drugs more widely consumed. Since the structures of cathinones are similar and their pk_a values are also very close, it is difficult to separate them by CE with sufficient resolution. Despite these difficulties, an interesting method has been published for the determination of twelve new psychoactive substances from the group of synthetic cathinones by MEKC-MS using a volatile BGE solution [62]. In order to minimize interferences with matrix components from urine samples and to pre-concentrate them, SPE was introduced as a clean-up step.

Table 2. Applications of CE-MS to the analysis of drugs of abuse

Analyte	Analytical technique	LODs	Type of sample	Reference
THC-COOH, THC-COOH-glucuronide	CE-MS	0.5 pg (10 nL injection)	Urine	[61]
DMA, *p*-hydroxymethamphetamine, AP, MA	CE-MS	0.03 - 0.05 μg/mL	Urine	[56]
AM, MA, MDA, MDMA, MOR, COD, COC, BZE, 6-AM	CE-ESI-ToF-MS	0.006 - 0.100 ng/mg	Hair	[66]
MDMA, MDA, 6-AM, AM, MA, MOR, COC, BZE, ephedrine, nalorphine (i.s.)	CZE-MS	0.005 - 0.100 ng/mg	Hair	[68]
Opiates (n = 14), amphetamines and derivatives (n = 10), COC and metabolites (n = 5), and others (n = 4)	CE-ToF-MS	2 - 200 ng/mL	Urine	[59]
Opiates (n = 12), amphetamines and derivatives (n = 10), COC and metabolites (n = 3), and others (n = 4)	DLLME-CE-ESI-TOF-MS	0.1 - 10 ng/mL	Urine	[60]
COC and metabolites (i.e., BZE, cocaethylene, anhydroecgonine, anhydroecgonine methyl ester and ecgonine methyl ester	CE-ESI-MS	100 - 200 ng/mL	Urine	[58]
EDDP, COD, hydrocodeine, 6-AM	In-line-SPE-CE-MS	0.013 - 0.21 ng/mL	Urine	[57]
MA, MDA, MDMA, MDA, MTD, COC, COD, BZE, MOR, 6-AM	CZE-ToF-MS	2 - 10 ng/mL	Blood	[64]
2,5-methylenedioxy-derivatives of 4-thio-phenylethylamine (n = 4)	CE-ESI-MS	11.3 - 23.0 ng/mL	Plasma	[65]
AM, MA, MDA, MDMA, MDEA, COC, COD, heroin, MOR	CE-ESI-MS	0.78 - 3.12 ng/mL	Urine	[63]
3,4-methylendioxypyrovalerone, buphedrone, ephedrone, naphyrone, methedrone, flephedrone, ethylone, cathinone, 3-fluormethcathinone, methylone, 4-methylethcathinone, mephedrone	SPE-MEKC-MS	10 - 78 ng/mL	Urine	[62]

On the other hand, many efforts have been done to carry out a correct hyphenation between the CEC mode and MS detection. In this sense, very few articles describing methods for the analysis of drugs of abuse by CEC-MS have been published. Aturki et al. have developed a laboratory-made liquid junction interface for the hyphenation of a CEC system with a MS detector for the analysis of nine illicit drugs in urine samples [63]. The coupling of packed CEC columns with MS using this interface provided several advantages such as better sensitivity, low dead volume and independent control of the conditions used for CEC separation and ESI analysis.

Despite urine analysis is widely implemented in forensic laboratories due to the non-invasive character of this matrix, blood is the specimen of choice for drug analysis in clinical and forensic toxicology because the drug concentration in blood can be directly correlated with the observed biological effect. An interesting method for the determination of numerous drugs of abuse in post-mortem blood samples collected from femoral veins has been proposed by Gottardo et al. [64]. Samples were extracted by LLE and injected under field-amplified sample stacking (FASS) conditions for achieving a sensitivity enhancement. In total, ten drugs of abuse including MA, MDA, MDMA, MDA, MTD, COC, COD, BZE, MOR and 6-AM were analyzed by CZE-ESI-ToF-MS within 25 min. On the other hand, the monitoring of phenethylamine designer drugs on the illicit drug market is a matter of concern for all authorities. Consequently, numerous clinical and forensic applications have been reported for its monitoring in human plasma. Under this context, Boatto et al. have developed a CE-ESI-MS method to screen and quantify four 2,5-methylenedioxy-derivatives of 4-thio-phenylethylamine in the so-called 2C-T-series in human plasma [65]. Prior to CE-MS, a simple liquid extraction was carried out for sample clean-up. Due to its simplicity, this method is susceptible to be applied to epidemiological and clinical studies.

Furthermore, in forensic toxicology, hair testing is widely employed to investigate drug chronic abuse. Despite the growing interest for this alternative toxicological sample, there are not many publications regarding the use CE-MS in the analysis of drugs of abuse in hair samples. The most important applications in this field have been developed by Gottardo et al. [66, 67, 68]. They developed the first CE-ESI-ToF-MS method for the accurate and efficient analysis of several drugs of abuse and their metabolites in hair [66]. Only a single LLE procedure was required for the extraction of all analytes (i.e., AM, MA, MDA, MDMA, ephedrine, COC, MOR, COD and their metabolites 6-AM and BZE), allowing the detection of a broad spectrum of basic drugs and their metabolites with moderate accuracy. In order to evaluate the applicability of the proposed method, hair samples collected from five volunteers following different drug rehabilitation programs [67]. This method demonstrated to be suitable for the rapid broad spectrum screening drugs of abuse in hair samples, although results must be further confirmed by molecular fragmentation in MS. Additionally, a CE-ESI-MS method using an ion trap (IT) spectrometer has been developed and validated for the rapid and simple determination of different amphetamines, ephedrine, COC and MOR in human hair in less than 20 min [68]. Hair samples from non-exposed subjects were analyzed for ensuring that matrix interferences were not present at the selected *m/z* values in the electropherograms. Interferences from other major drugs of abuse (i.e., barbiturates, THC and its acid metabolite, BZDs) were also excluded.

Table 2 shows the most relevant CE-MS methods developed for the determination of drugs of abuse in different forensic matrices.

Other Modes of Detection

As indicated above, CE-UV coupling usually does not provide the detection limit that is required for the analysis of biological samples in forensic toxicology. Consequently, the application of pre-concentration techniques is usually needed for overcoming the low concentration sensitivity. Another possibility to increase sensitivity is the use of luminescence detectors [69].

Fluorescence spectroscopy has been widely used in many fields of science, but there are few applications in the field of forensic analysis. This is due to the fact that the coupling of CE instrumentation with fluorescence or LIF detection is instrumentally complex as well as few drugs of abuse show native fluorescence. In the case of LIF, specific laser sources are also required depending on the analytes. At the moment, few papers have been found using CE coupled with LIF.

Several drugs such as normorphine, 6-AM, MOR, and COD possess native fluorescence. Consequently, a CE-fluorescence method was developed for the separation and determination without requiring any derivatization procedure, and using an excitation wavelength of 245 nm with a cut-off emission filter of 320 nm [70]. The detection limits were in the range of 200 ng/mL. To increase the sensitivity of the present analysis, LIF detection was also employed using a precolumn derivatization procedure with the most frequently used derivatizating agent FITC which lower the LODs to the range between 50 - 100 pg/mL.

One of the first applications of CE-LIF for forensic analysis was intended for the determination of the allucinogenic drug LSD, taking advantage of its native fluorescence, in blood samples [71]. Analysis time was of 8 min. Using a specific sample preparation, electrokinetic injection and a bubble cell, LODs of 0.1 – 0.2 ng/mL were obtained with an accuracy between 86 – 110%. This method was applied to human blood samples from two forensic cases. Results were consistent with those achieved by a radioimmunoassay.

The determination of a mixture of ephedrine and amphetamine drugs has been carried out by a CE method using pre-column derivatization and LIF detection [72]. Following derivatization with FITC, a total of six drug of abuse (i.e., AM, MA, MDA, MDMA, pseudoephedrine and ephedrine) as well as propafenone as internal standard were analyzed within 20 min. The method was validated in human blood and urine samples, reaching LODs of 0.2 ng/mL. Amines can be easily derivatized with FITC and analyzed by CE using alkaline buffers with or without SDS. Both CZE and MEKC coupled with LIF have been evaluated for the analysis of four amphetamines (AM, MA, MDMA and β-ohenylethylamine) in human urine using chip-based and fused-silica capillary instrumentation [73].

Electrochemical detectors, especially those based on amperometry detection, are more sensitive than UV detectors and cheaper than LIF and MS detectors, and provide excellent sensitivity and high selectivity towards electroactive drugs. Capacitively coupled contactless conductivity detection (C^4D) is a universal and low-cost detection system that has been applied in many different analytical fields [74]. In this sense, CE-C^4D has also been applied to the analysis of drugs of abuse, specifically, to the determination of seven amphetamine analogues as well as AM, MA, dextroamphetamine, and MDMA [75]. CE-C^4D has also been employed for the enantiomeric separation of different stimulants (i.e., AM, MA, EPH, pseudoephedrine, norephedrine and norpseudoephedrine) [76]. In order to achieve the enantiomeric separation of ephedrine and pseudoephedrine, two chiral selectors, namely heptakis(2,6-di-*O*-methyl)-β-cyclodextrin (DMBCD) and carboxymethyl-β-cyclodextrin (CMBCD), and a modifier (hydroxypropyl) methyl cellulose (HPMC) were added to the BGE solution. LODs ranging between 2.3 and 5.7 μmol/L were obtained, with good precision in migration times and peak area.

Some of the most common illegal drugs such as GHB have a very short half-life, around 20 – 50 min, being rapidly eliminated from the body. Therefore, the detection window is rather short. Thus, samples should be collected as quickly as possible after intake and, especially, oral fluids such as saliva. An interesting application has been developed by Mazina et al. for the determination of GHB in saliva by CE combined with C^4D and indirect UV detection at 210 nm [77]. Instrumental LODs were 0.49 and 5.1 mg/L for C^4D and indirect UV detection, respectively. Both methodologies offer a rapid, accurate and sensitive analysis for the determination of this well-known and popular recreational drug, which is also used as drug for facilitating sexual assault, robbery, money extortion among other crimes. Furthermore, CE-C^4D was earlier applied to the analysis of GHB in other biological samples such as urine and serum. Simultaneous separation of GHB and β-hydroxybutiric acid, which is also present in clinical samples, was directly carried out since only a four-fold dilution was required prior analysis [78]. LODs in the clinical samples were found to be about 2 μg/mL, providing enough sensitivity for the recognition of over dosage.

Recently, CE-C^4D has been proposed for the simultaneous determination of GHB in presence of a neurotoxin (i.e., ibotenic acid) and a hallucinogen (i.e., psilocybin) in human saliva [79]. These three substances can be lethal, especially if they are consumed by accident or in an unknown dosage. LODs and limits of quantification (LOQs) in saliva were 1.5 and 5.0 mg/L for GHB, 2.1 and 7.9 mg/L for ibotenic acid and 3.6 and 12.0 mg/L for psilocybin. This method showed that the use of oral fluid as an alternative sample matrix, which is easy and fast to collect and prepare, in combination with a portable CE instrument provides a rapid and efficient determination of three important psychoactive substances.

In general, CE coupled with conductivity detection is a rapid methodology for the determination of small anionic species in different biological samples such as blood, urine or saliva, which is very appropriate for cases where the substance of abuse is unknown and a rapid detection is needed.

FORENSIC DNA

Nowadays, deoxyribonucleic acid (DNA) analysis is fundamental in forensic investigations [80]. Human genome sequences are nearly 99% identical, but two individuals can be unequivocally discriminated based on DNA polymorphisms. In other words, the sequence of DNA is characteristic of each person, except in the case of monozygotic/identical twins who share DNA fingerprints. There exists a very low probability (typically ~10^{-18}–10^{-25}) that two random individuals have the same genotype (DNA profile) by chance [81]. Consequently, DNA evidence can link the suspect with the victim and/or with the crime scene, providing a solid base for the conviction of the suspect. It is irrefutable evidence in many criminal investigations such as, for example, in the identification of a rape suspect. Perhaps more importantly, DNA testing has led to reopen many old cases and proved that innocent people were wrongly convicted before the benefit of DNA evidence was introduced, as shown by Innocence Project. DNA analysis is also required in paternity suits, in cases of missing people, and for the identification of victims of natural disasters or unknown soldiers killed in combat.

In the mid-eighties, at the University of Leicester (United Kingdom, UK), a team of researchers headed by Professor Alec Jeffreys developed the first technique for producing DNA fingerprints from DNA samples [82, 83]. DNA is mainly located in the nucleus of the cells and, in forensic investigations, blood and semen are the main sources of DNA. As result of current advances in science and technology, DNA evidence can also be collected from other body fluids and tissues (e.g., saliva, sweat, hair, bones, etc.) as well as from touched objects [84]. DNA profiling was applied for the first time in a forensic investigation in 1986. It led to the identification of Colin Pitchfork as the murderer and rapist of two girls in the English Midlands (UK) and to the exoneration of a 17 year-old boy who incriminated himself [85]. Since then, DNA analysis has replaced other methods used for typing blood and semen stains such as those involving blood group markers or protein polymorphisms. The book *Fundamentals of Forensic DNA Typing* gives a complete overview of the history and current state of DNA typing [86].

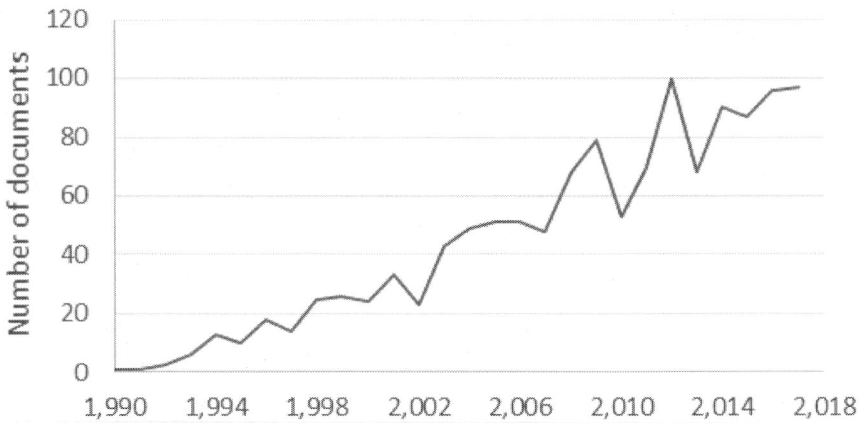

Figure 2. Number of publications (only research articles) since 1990. Search was carried out with Scopus database using the terms 'forensic DNA' and 'capillary electrophoresis' as keywords.

Polyacrylamide gel electrophoresis (PAGE) and agarose slab gel electrophoresis has been traditionally used for the analysis of DNA fragments. However, CE using molecular sieving separation modes quickly became the method of choice for DNA typing due to its high efficiency and low analysis time [87, 88]. As indicated in the series of articles of *Electrophoresis* related to CE applications in forensic analysis, CGE coupled with fluorescence detection is considered the 'gold standard technique' for DNA fingerprinting and sequencing [89, 90, 91, 92]. DNA analysis by CE is currently accepted in American courts of law as well as in many other countries [93]. Under this context, the number of research articles dealing with 'forensic DNA analysis' and 'capillary electrophoresis' has grown significantly in the last years. As observed in Figure 2, nearly 100 documents involving both terms are annually published. Therefore, it is almost impossible to present a comprehensive review of the scientific production generated in the area. In addition, most of the research articles are mainly focused on particular applications and the CE-related methods are not significantly discussed. This section provides a holistic view of current DNA typing strategies, in which CE is involved, and also describes how the 'state of the art' of CE in DNA analysis has evolved in the last decades.

DNA Typing Strategies

In forensic investigations, only small sections of the DNA sequence are examined, but they provide enough information to discriminate between two DNA samples with high confidence. Short tandem repeats (STRs), also known as micro-satellites or simply as DNA sequence repeats, constitute DNA segments in which a pattern of two to six base pairs (bp) is continuously repeated in a row on the DNA strand. STR markers are highly useful for human identification purposes because it is normal that two individuals present different number of repeating units (i.e., alleles) at a given position (i.e., locus) on a chromosome. On the contrary, they can also have the same allele at the same STR locus purely by chance. For that reason, several STR loci must be analyzed in order to discard any false positive when matching DNA fingerprints. As a result of scientific and technological advances, the number of examined STR loci has increased over time. Since 2014, 16 loci are analyzed in investigations conducted within the UK whereas 23 loci are examined in some Scottish cases [94]. DNA databases based on STRs are well established in more than 60 countries, of which the most important are CODIS (i.e., the American DNA database created and maintained by FBI) and NDNAD (i.e., the UK National DNA Database) [95, 96]. According to CODIS, 13 STR markers have been traditionally examined in criminal investigations carried out by FBI. Since January 2017, the evaluation of 7 additional STR loci is required in order to reduce the number of adventitious matches and increase the compatibility with international databases [97].

At present, STR fingerprinting is the most widely used analytical approach for DNA-based human identification. It offers extremely high discrimination power in comparison with other techniques [98]. In forensic DNA laboratories, DNA samples are firstly submitted to polymerase chain reaction (PCR) in order to amplify or copy segments of DNA and, consequently, improve sensitivity. As a result, DNA typing by STR-based methodology can be accomplished with extremely small amounts of DNA. PCR products, which must be fluorescently labeled, are subsequently analyzed by CE-LIF.

The most commonly used CE systems for STR analysis are the single capillary ABI Prism 310 Genetic Analyzer and the multi-capillary ABI 3100 or 3130xl (Applied Biosystems, Foster City, CA). Since 2010, the 8-capillary ABI 3500 and 24-capillary 3500xl instruments are also available with new capabilities [99]. Butler et al. have reviewed the most relevant aspects of both the ABI Prism 310 and 3100 genetic analyzers for the analysis of STR markers [100]. Different commercial kits are available for the simultaneous amplification of several STR markers and their subsequent analysis in one or two electrophoretic runs. CE methods using this technology have been rigorously validated, showing that they are sufficiently robust for their implementation into forensic laboratories [101, 102, 103].

On the other hand, a wide range of samples can be collected as forensic evidence for STR typing. In addition to common forensic samples (i.e., body fluids and tissues), personal effects such as cosmetic applicators (e.g., lipstick) or toothbrush can also be an important source of DNA [104, 105].

Despite its great advantages, STR profiling presents certain limitations. The typing of degraded DNA samples can hardly be carried out by STR-based approaches [98]. The analysis of DNA mixtures can also result a challenge [97]. Consequently, other methods involving the analysis of single nucleotide polymorphisms (SNPs), Y-chromosome STRs (Y-STRs) or mitochondrial DNA (mtDNA), have been developed [106].

SNPs represent the most abundant class of human polymorphisms and can be analyzed in short amplicons (i.e., amplified DNA fragments), which results highly useful for the fingerprinting of degraded DNA samples [107]. Only ~50 bp are needed for SNP typing whereas ~300 bp are required to successfully accomplish STR profiling [98]. Sanchez et al. have identified and proposed a set of 52 SNPs for human identification [108]. DNA fragments were submitted to PCR amplification and subsequently analyzed by CE using ABI Prism 310, ABI Prism 3100 and ABI Prism 3100-Avant Genetic Analyzers with 36 cm capillary arrays. Amplicons length ranged between 59 and 115 bp. The discrimination power of this DNA typing approach (> 99.9999%) was calculated using a sample group of 700 individuals from Denmark, Greenland, Somalia, Turkey, China, Germany, Taiwan, Thailand and Japan. This SNP typing method led to a combined mean match probability of 5.0×10^{-19} and 5×10^{-21} for Asian and European populations, respectively. It was also observed that two unrelated individuals did not exhibit more than 41 identical loci within the 52 loci evaluated. Nevertheless, there is a general consensus that SNP typing will not replace STR-based methods for now.

The examination of STR-loci located on the male Y-chromosome has emerged as an alternative methodology for DNA analysis. Due to its specificity, it is a powerful strategy for DNA profiling in the case of DNA mixtures with male and female contributors (e.g., DNA evidence collected in rape investigations where male perpetrator's DNA is mixed with female victim's DNA). Y-STR typing is effective even when female DNA exceeds 4000 times the amount of male DNA [109]. Nevertheless, Y-STRs are permanently linked and mutations are rarely observed between close patrilineal relatives, so this approach provides lower discrimination power than STR typing [81]. In this sense, great efforts are dedicated to the discovery of characteristic Y-STRs for enhancing the discrimination power of this technique. A recent study has shown that the fingerprinting of certain rapidly mutating (RM) Y-STRs allows to distinguish between DNA samples from father and son in 50% of cases, and from brothers in 60% of cases [110]. In this study, CE (i.e., ABI3130xl Genetic Analyzer) was used as analytical technique.

In addition to the investigation of nuclear DNA, the DNA contained in the mitochondria of the cells can also be examined for DNA typing [111]. The mtDNA is maternally inherited. Moreover, mtDNA is naturally more abundant and is degraded at lower rate than nuclear DNA. The main disadvantage of mtDNA profiling is the same as of Y-STR typing, because mtDNA is non-recombining and mutations are not normally observed between individuals from the same maternal lineage [98]. In general, the analysis of mtDNA is usually applied in the case of degraded DNA samples or where nuclear DNA may not be present, including skeletal remains, shed hair or hair fragments, teeth, etc. In 2000, the remains of Mary Sullivan, who was the thirteenth and last victim of Boston Strangler (United States of America, USA) in the early 1960s, were exhumed and analyzed in order to confirm the culpability of the self-confessed perpetrator, Albert DeSalvo [112]. DNA sequencing was based on mtDNA samples collected from her underwear and pubic hair. DNA samples were amplified by PCR and subsequently analyzed with an ABI 310 genetic analyzer. Two different mtDNA sequences were generated and neither of them matched the victim nor members of the forensic team involved in the investigation. Moreover, none of the two samples was attributed to Albert DeSalvo, who was never a crime suspect and was murdered in prison in 1973. Nowadays, there is still controversy about the real identity of Boston Strangler.

In addition to DNA typing approaches for human identification, the identification of cell types contained in biological stains and found at the crime scene may be crucial for the forensic investigation. In order to typify the biological origin of forensic evidence, several methodologies based on different molecular markers, namely messenger ribonucleic acid (RNA), micro-RNA, DNA methylation and microbial markers, have been developed in the last years [113]. Among these techniques, mRNA typing is currently widely applied. The mRNA constitutes a family of RNA molecules that convey the DNA code from the cell nucleus to the sites of protein synthesis in the cytoplasm (i.e., ribosomes). Different mRNA markers can be used for the identification of body fluids such as blood, semen, saliva, vaginal secretion and menstrual blood as well as skin [114, 115]. From the analytical point of view, mRNA profiling typically involves sample extraction, PCR amplification, CE separation and detection by LIF.

State of the Art of CE in DNA Analysis

Unlike in other forensic fields, CE is well implemented in forensic DNA laboratories. CZE is the most commonly used CE mode and has been applied to the separation of DNA fragments, but it is ineffective for the analysis of DNA fragments with different length and similar charge-to-size ratio. In general, CGE is the most suitable separation mode for DNA analysis [116, 117]. In CGE, capillaries are filled with a sieving gel that allows the separation of DNA fragments based on their molecular size. Smaller DNA molecules move easily through the network of the gel and, as a consequence, migrate faster than larger molecules. DNA separations are normally carried out in coated capillaries under reverse phase conditions (i.e., cathode located at the inlet end of the capillary). Typically, DNA samples are electrokinetically injected since they are negatively charged molecules in neutral pH media. Buffers used in DNA separations usually have a pH around 8. Under this condition, the presence of EOF is highly relevant, so capillaries are coated in order to suppress it [99]. Different polymers such as PVA, poly(dimethylacrylamide) (PDMA), etc. can be applied as capillary coating for the analysis of DNA fragments [118]. More detailed information about DNA sample preparation, sample injection, and buffer characteristics can be found in specific literature reviews and book chapters [99, 100].

As a first attempt, gel-filled capillaries were evaluated as sieving medium in CGE for DNA sequencing. Several disadvantages are attributed to the use of gel-filled capillaries such as the formation of air bubbles during their production process and their short lifetimes. For these reasons, they are scarcely employed in DNA separations [100, 117]. Nowadays, CGE matrices consist of polymers and, within them, the use of linear polyacrylamide (LPA) is widespread. LPA offers great performance at low cost but, due to its high viscosity and lack of capability to act as capillary coating, other alternative polymers are also used. Indeed, the use of matrices with coating properties is highly required because the capillary lifetime is extended [116, 119]. PDMA, hydroxyethlycellulose (HEC), polyvinyl pyrrolidone (PVP), etc. have also been evaluated as DNA sieving matrices [120]. PDMA is a low-viscosity matrix and one of the most widely used matrices in DNA sequencing. The users of ABI's analyzers generally use capillaries consisting of polyolefin plastomer polymers (POPs, i.e., POP-4™, POP-6™, and POP-7™), which are PDMA-based polymers [119]. Despite this, different materials have been evaluated in order to improve the performance characteristics of genotyping matrices and, consequently, DNA separations. In this sense, the combination of polymers has been widely explored [121, 122]. Boulos et al. have developed a sieving matrix consisting of mixture of PVP and HEC (20.4% of PVP relative to HEC) [123]. This low-viscosity matrix combines the 'self-coating' properties of PVP and the high efficiency of HEC. It has shown long lifetime (more than 90 successive runs were carried out without requiring capillary conditioning between runs) and high separation performance in terms of repeatability (RSDs of 2.2 and 5.3% for the separation of DNA fragments of 139/150 and 490/500 bp, respectively).

Special attention must be paid to coating and sieving matrix materials when microdevices are used for DNA typing. Materials typically used for DNA typing in CGE systems may not be the most optimal for DNA sequencing in miniaturized systems such as microchips [124]. Nevertheless, a wide range of materials are currently available for the production of microchips [125]. Microchips are generally designed for one-time use, so cheap materials are required in order to avoid high analysis costs. STR profiling using a microdevice can be ten times more expensive than using conventional methods [126].

On the other hand, particular mention should be made of the technological development undergone by instrumentation in the field of DNA analysis. Undoubtedly, the development of multi-capillary devices, also referred as capillary array electrophoresis (CAE) instruments, has increased the throughput of forensic laboratories. In this regard, CAE technology has been key to sequence the Human Genome [127]. In addition to traditional commercial instruments, multiplexed CE-LIF systems are under continuous development [128, 129], giving rise to numerous patents related to this technology. Within this framework, an interesting trend is the development of microfabricated capillary array electrophoresis (μCAE) devices for DNA analysis. Microfluidic devices consume lower amount of reagents than conventional CAE instruments and offer higher efficiency and shorter analysis time. In comparison with CAE instruments, μCAE devices increase the sequencing rate by five- to ten-fold [130]. In general, DNA typing in microchip-based capillary electrophoresis (μCE) systems can be almost two times quicker than in traditional CE instruments [131, 132]. Yeung et al. have proposed a 96-microchannel μCAE system for STR typing in semen, saliva and blood, with the goal of increasing DNA analysis throughput [133]. This device has demonstrated to be 10 min faster than the single-capillary ABI 310 and 30 min faster than the 16-capillary 3100 instruments, providing similar sensitivity and capacity to detect minor STR alleles.

It is very important to note that μCE technology has already been transferred from laboratory scale to industry scale. Several μCE devices are currently commercially available (e.g., Agilent Technologies' 5100 automated Lab-on-a-Chip platform, Lab Chip 90 system from Caliper, etc.), which may contribute to the implementation of this technology in forensic DNA analysis [134]. Furthermore, one of the most significant advantages of μCAE systems is that they can be directly applied at the crime scene reducing the risk of contamination [135]. For this to happen, the portable device must integrate not only the μCE system, but also DNA extraction and PCR amplification.

A wide range of microfluidic systems, also referred as lab-on-a-chip devices, has been proposed for DNA analysis in the last years. [125, 135, 136, 137]. In these microfluidic systems, all processes involved in DNA typing (i.e., DNA extraction, PCR amplification and electrophoretic separation) can be carried out in a partially integrated microdevice (e.g., a micro-solid phase extraction (μSPE)–micro-PCR (μPCR) chip, a μPCR–μCE chip) or in a fully integrated system (μSPE-μPCR-μCE) [138].

Moreover, integrated lab-on-chip devices can consist of one single microchip or a fully integrated device where each process occurs in an independent microchip. For example, Le Roux et al. have recently developed an integrated microchip that combines multiplex PCR amplification and electrophoretic separation for STR typing of forensic samples [132]. This microfluidic device was fabricated in inexpensive cyclic olefin polymer (COP) plastic and contains four identical microchannels. As a result, it could allow the simultaneous analysis of four samples. Each 7-cm long microchannel was filled with the sieving matrix consisting of 4.5% (w/v) hydrophobically modified poly(acrylamide) (HMPAM). In comparison with conventional approaches, STR profiling time could be reduced from more than 2.5 hours to less than 90 min by using the developed microchip. On the other hand, Han et al. have developed an integrated microdevice that combines two plastic microchips for DNA extraction and PCR amplification with a glass μCAE for automated STR typing [126]. DNA extraction and amplification chips consisted of poly(methyl methacrylate) (PMMA) and were of one-time use, whereas glass μCAE could be reused after cleaning with deionized water. CAE chip contained two separation channels of 16 cm long that were filled with 4.5% (w/v) LPA as separation medium. In addition to its low cost, this microdevice allowed STR profiling from buccal swab and whole blood samples, enhancing the applicability of microfluidic systems which are mainly designed for DNA typing from buccal swabs.

The applicability of a portable microchip analyzer for STR typing at the crime scene has been recently demonstrated by Liu et al. [139]. The analyzer integrated a PCR-CE microchip and four-color fluorescence detection. The microchip was fabricated in PDMS (polydimethysiloxane) and contained a 160-nL PCR reactor and a 7-cm long CE separation channel. For DNA fingerprinting, a 9-plex autosomal STR typing system was developed and included amelogenin, a sex-typing marker, and 8 STR loci (D3S1358, THO1, D21S11, D5S818, D13S317, D7S820, vWA, and D8S1179) used by CODIS. The portable analyzer was applied in a mock crime scene prepared by the Palm Beach County Sheriff's Office (FL, USA). The entire forensic process only took 6 hours, including evidence collection, DNA extraction, PCR amplification, CE separation, generation of DNA profile and search againist the mock CODIS convicted offender database. On the contrary, conventional STR typing approaches require at least of 3 days from sample collection to CODIS hit generation. In short, real-time forensic DNA analysis represents progress for the rapid identification of a suspect after the commitment of a crime.

In addition to μCE-based systems for forensic STR profiling, lab-on-chip devices have also been proposed for accomplishing other DNA typing strategies such as Y-STR [140, 141] or mtDNA [142].

EXPLOSIVE COMPOUNDS AND GUNSHOT RESIDUES

Explosive compounds employed for criminal and terrorism activities, and GSRs that can be found on the skin or clothing of people who have fired a weapon, on the victim or on other targeted materials at the crime scene, may provide valuable evidence in forensic investigations. However, explosive compounds or GSRs may be present at very low or trace amounts and they are usually found in very complex matrices. For these reasons, the interpretation of their analysis is a challenge for forensic chemists.

Explosives include a great variety of compounds. They are usually divided into two broad categories: high and low explosives. High explosives can be further subdivided into primary explosives, which burn very quickly or detonate when ignited, and are extremely sensitive to heat, shock, or friction (e.g., nitroglycerin (NG), acetone peroxide, tetrazene, etc.), and secondary explosives, which are much less sensitive, require substantially more energy to be initiated, and are typically initiated via a smaller quantity of a primary explosive (e.g., cyclotrimethylenetrinitramine (RDX) or trinitrotoluene (TNT)). Low explosives, also called homemade explosives (HMEs), are based on inorganic and organic compounds that can lead to the formation of the so-called peroxide based explosives, which can be both primary and secondary explosives.

The analysis of explosive compounds can be oriented to the identification of these substances prior or after their detonation, known as pre- and post-blast analysis, respectively. Post-blast explosive analysis is the most frequently one and refers to the analysis of explosive compounds after a bombing incident, based on the fact that not all the explosive materials are consumed during the explosion. In this sense, it would also be necessary to take into account the by-products that can be formed by chemical reactions occurred during the explosion.

As can be seen, explosive analysis is a difficult task that needs a wide range of analytical techniques with high requirements in terms of speed, selectivity, sensitivity and robustness. For many years, GC has been used for the identification of organic explosive compounds because it provides high selectivity and sensitivity for the analysis of organic compounds. However, the application of GC is limited to volatile substances. In general, different detection systems can be used when GC is selected as separation technique, including electron capture detectors (ECDs), thermal energy analyzers (TEAs), and mass spectrometers. HPLC is a complementary technique to GC that can be applied to the analysis of non-volatile, low volatile, high sensitivity to heat and high molecular weight organic explosives. Nowadays, ultra-high performance liquid chromatography (UHPLC) is the analytical technique of choice for the analysis of non-volatile compounds because it provides high separation efficiency in short analysis times. Additionally, UHPLC coupled with MS or MS/MS is often applied due to its high selectivity and sensitivity.

The analysis of inorganic low explosives is also of great interest to forensic investigators. In most cases, anionic and cationic compounds are usually determined separately using different separation systems with different types of columns and conditions. In this sense, ion chromatography (IC) has been widely employed in the analysis of explosives to separate both inorganic anions and cations found in post-blast residues.

In addition, there are other instrumental techniques that are also used for the analysis of explosives such as infrared spectroscopy (IR), X-ray powder diffraction (XRD) and X-ray fluorescence (XRF). These techniques are non-destructive, so their application is widely recommended for preserving the integrity of evidence.

All of the abovementioned instrumental techniques are used in combination with presumptive tests to facilitate, in the first place, the identification of explosive residues. Spot test kits and thin layer chromatography (TLC) are examples of preliminary color tests that are often employed in forensic laboratories for the screening of explosive residues. Furthermore, X-ray imaging detectors as well as spectroscopic sensors are widely used in security posts, airports and border controls in order to avoid that prohibited items are passed through the security line.

In the field of explosive analysis, CE serves as a complimentary technique to IC for the analysis of inorganic low explosive residues. CE is ideal for the analysis of these residues because it requires low amount of sample and can be applied to both anions and cations. For many years now, CE has demonstrated to offer a valuable alternative to other well-established instrumental techniques in the field of forensic analysis, due to its low running costs and simplicity of use [143, 144].

Regarding GSRs, they are trace evidence resulted from the criminal use of firearms and can link the shooter with the firearm, the victim and/or the crime scene [145]. Due to the diversity of GSRs, the classification recently carried out by Goudsmits et al. is very valuable [146]. They have classified this type of compounds in three categories, covering more than 136 substances that can be associated with organic GSRs.

The analysis of GSRs is one of the most important tasks of forensic science. Traditional techniques lack of specificity and are mainly focused on the determination of inorganic elements mainly lead, barium, and antimony, although other less common elements can also be analyzed (e.g., aluminum, sulfur, tin, calcium, potassium, chlorine, copper, strontium, zinc, titanium, or silicon). Nevertheless, modern gunpowder, also called smokeless powder, contains organic compounds, which primarily comes from materials derived from the propellant and may be classified either as explosives or additives, depending on their chemical composition.

The most common explosive found in GSRs is NG whereas the additives may include stabilizers, plasticizers, moderants, coolants, flash inhibitors and other components intended to improve the performance of the powder. For a correct GSRs analysis, it is very important to obtain the samples from the skin surfaces of the victim at the scene as rapid as possible. Ideally, sample collection should be carried out from 1-3 hours after the discharge of the firearm discharge to maximum 48 hours in order to obtain the maximum sample recoveries. Any difficult or delay in obtaining these residues as well as the movement or washing of the body prior to autopsy will diminish or destroy this type of evidence. As it can be deduced for the complexity of the compounds that can be found in GSRs, their identification requires robust analytical methods.

Several techniques of elemental analysis have been used for the determination of GSRs, including color reactions, neutron activation analysis (NAA), electrothermal atomic absorption spectroscopy (ETAAS) and inductively coupled plasma-mass spectrometry (ICP-MS). Nevertheless, due to the fact that scanning electron microscopy-energy dispersive X-ray spectrometry (SEM-EDX) combines the morphological and elemental analysis of metal residues, it has been one of the most widely accepted techniques for the accomplishment of this purpose. However, the modern trend to produce primers free of heavy metals has reduced its application because, in these cases, it may lead false negative results [147].

Traditionally, inorganic residues have been analyzed by the elemental techniques mentioned above while organic GSRs have been analyzed mainly by chromatographic techniques using different detection systems. Nevertheless, CE can be employed for the simultaneous determination of organic and inorganic compounds. Moreover, it is cheaper, faster and more specific than other separation techniques. For these simultaneous determinations, and prior to CE analysis, complexation of metal ions must be carried out. In these cases, MEKC is used as separation mode, which is also suitable for uncharged organic compounds. Other GSRs and heavy metals can also be analyzed by CE. Consequently, this technique has been accepted in forensic laboratories [148, 149].

The aim of this section is to report the current applications of CE for the analysis of two important types of evidence such as explosive compounds and GSRs.

CE for the Analysis of Explosives Compounds

The analysis of low inorganic explosives has been traditionally carried out by IC. Nevertheless, due to its inherent characteristics of shorter analysis times with the possibility of being applied to both anions and cations, CE has demonstrated to serve as a complimentary technique to IC since the work of McCord in 1992 [150]. However, the low sensitivity achieved by CE, as consequence of the low amount of analytes that are injected, still remains as its main disadvantage.

Hopper et al. developed the first CZE method using indirect UV detection for the simultaneous separation of anions and cations, and the further analysis of a variety of low explosives in post-blast residues [151]. Both anions (i.e., Cl^-, NO_2^-, NO_3^-, SO_4^{2-}, ClO_4^-, SCN^-, ClO_3^-, and OCN^-) and cations (i.e., NH_4^+, K^+, Na^+, Ca^{2+}, and Mg^{2+}) were successfully separated in a single run in less than 7 min. Additionally, Ahmad et al. developed a CE method for the analysis of inorganic ions using 25 mM 2,6-pyridinedicarboxylic acid (PDC) as a BGE (pH 4.7) and 0.5 mM cetryltrimethylammonium hydroxide (CTAH) to reverse the direction of the EOF [152]. This method was successfully applied to the simultaneous separation of different anions (i.e., Cl^-, NO_3^-, SO_4^{2-}, and SCN^-) and cations (i.e., Ca^{2+}, Fe^{2+}, and Fe^{3+}) in 7 min. LODs ranged from 0.5 to 15 mg/L and the method was applied to the analysis of different post-blast residues. Another CZE with indirect UV detection has also been developed for the simultaneous determination of five anions (i.e., Cl^-, ClO_4^-, NO_2^-, NO_3^-, and SO_4^{2-}) and two cations (i.e., NH_4^+, and K^+) [153]. All anions and cations were separated in less than 4.6 min and 2.6 min, respectively. The method was successfully applied to real samples. Moreover, different matrices such as soil, metal, plastic, paper, cotton and clothes have been evaluated for the analysis of several anions (i.e., Cl^-, NO_2^-, NO_3^-, $S_2O_3^{2-}$, ClO_4^-, SCN^-, N_3^-, ClO_3^-, SO_4^{2-}, HCO_3^-, HPO_4^{2-}, formate (internal standard, i.s.), I^-, Br^-, F^-, OCN^-, acetate, oxalate, and chlorite) in pre- and post-blast residues [154]. Analyses were carried out in less than 20 min by CE-UV. The results obtained by CE were compared with those achivied by an IC method used in routine analysis, showing excellent correlation. CE has also demonstrated to be a powerful technique to detect and quantify inorganic species and carbohydrates in post-blast residue extracts [155]. Sucrose, the main component of icing sugar, was present at a concentration of 145 mg/L. In addition, fructose and glucose, which probably resulted from the degradation of sucrose, were detected at 11 and 25 mg/L, respectively. This fact confirmed the use of this carbohydrate in the explosive charge.

On the other hand, the first known application of miniaturized isotachophoresis (ITP) for the analysis of inorganic explosive residues was developed in 2008 [156]. This CE mode provides a number of useful features compared to CZE such as the ability to pre-concentrate diluted samples and the ability to limit the species that can be detected by carefully selecting the electrolytes employed. The proposed ITP-based method offered the advantage of being able to simultaneously analyze chloride, chlorate and perchlorate, which is a demanding analytical challenge these species present similar electrophoretic mobilities. This finding is especially important due to the fact that previous IC methods were not able to achieve their separation.

Improvised explosive devices (IEDs) are based on combinations of strong inorganic oxidizers and fuels and, in contrast to the strictly controlled organic high explosives, inorganic IED components can be easily and legally obtained at low cost. This underlines the importance of their analysis and detection. A rapid separation of relevant anions (i.e., Cl^-, NO_3^-, N_3^-, ClO_4^-, ClO_3^-, SCN^-, CO_3^{2-}, SO_4^{2-}, F^-, and PO_4^{3-}) to the identification of IEDs have been carried out by a simple SI-CE instrument coupled to C^4D [157]. The detector response for all analytes was linear over the range of 0.025 to 3 mg/L and the applicability of the method was demonstrated by the analysis of soil extracts collected after the controlled detonation of inorganic IEDs. In addition, a commercial portable CE instrument with a LED detector has been employed to enable a sensitive indirect photometric detection of 15 anions (i.e., chloride, thiosulfate, nitrite, nitrate, sulfate, perchlorate, thiocyanate, chlorate, cyanate, fluoride, chlorite, phosphate, carbonate, acetate and benzoate) and 12 cations (i.e., monomethylammonium, ammonium, ethylammonium, potassium, sodium, barium, strontium, magnesium, calcium, manganese, zinc and lead) found in post-blast residues from IEDs [158].

Cellulose nitrate, commonly known as nitrocellulose (NC), is employed in the manufacturing of explosive materials such as smokeless gunpowder and dynamites. For such a reason, methods for the determination of nitrogen content in NCs are needed. In 2014 and for the first time, CE was used to quantify nitrite and nitrate ions released after alkaline hydrolysis of NC in order to determine its nitrogen content in real gunpowder [159].

In general, UV/Vis detection is commonly used in CE methods intended for the analysis of explosives, but sometimes it is lack of sufficient sensitivity. Electrochemical methods and specially conductivity detection is the most universal form of detection for inorganic ions and, in the case of CE separations, C^4D has been used with increasing frequency over recent years. Within this framework, a portable CE-C^4D instrument has been evaluated for the simultaneous analysis of a wide range of anions a cations, reaching very low LODs in samples of pre- and post-blast residues [160]. Another method was published using the same detection but carrying out the separation on a CE-based chip, which permitted the analysis of a mixture of seven explosive-related cations and anions [161]. This type of microsystems offers great promise for monitoring explosive-related ions at the sample source, with significant advantages of speed/warning, efficiency, cost, or sample size.

The benefit of simultaneous analysis of anions and cations is clear. With this goal in mind, a SI-CE-C^4D method was developed for the rapid separation and identification of several inorganic anions and cations based on a single sample injection [162]. Various environmental and industrial samples were analyzed in less than 3 min, only subjected to filtering, and allowing the separation of inorganic and small organic anions and cations with LODs ranging from 16 to 61 μg/L.

Both organic explosives and peroxide-based explosives are neutral compounds that are mainly separated by the MEKC. The first MEKC application for the analysis of this type of explosives was reported by Northtop et al. in 1991, in the context of the investigation of 26 gunpowder and organic explosive constituents [163]. A MEKC method using UV detection at 200 nm was also applied to the analysis of these type of explosives by Cameron et al. [164]. The simultaneous separation of 25 organic explosives was carried out in 17 min, without requiring any sample pretreatment and with a very high efficiency. In addition to MEKC, this type of explosives has also been analyzed by CEC [165, 166, 167]. Bailey et al. evaluated both modes, MEKC and CEC, for the simultaneous analysis of organic explosive compounds using LIF detection [165]. However, the main limitation of this detection mode is that explosive compounds do not have significant fluorescence yields. Therefore, in order to enhance sensitivity, indirect LIF detection was employed. According to the authors, it supposes the first example of indirect LIF detection in packed CEC columns. Furthermore, in both CEC and MEKC, the high running currents created thermal instabilities, causing fluctuations in the background and limiting the LOD. This effect was worse in MEKC.

The wide availability of smokeless powders makes them particularly interesting as primary charge in IEDs. In an attempt to improve their detection, De Perre et al. evaluated the analysis of different explosives by CEC-UV-Visible and CEC-ToF-MS [168]. The goal was to classify smokeless powders on the basis of the various components present in the additive packages combined with the nitrocellulose base. The high resolution provided by the monolithic CEC columns allowed the separation of the isomeric forms of DNT, whereas the high mass accuracy afforded by ToF-MS provided the unambiguous determination of many smokeless powder components.

CE for the Analysis of Gunshot Residues

GSRs may be left at the crime scene after the shooting of a firearm. They can consist of both burned and unburned particles from the original explosive, and can be in presence of additional residues from the surface of the bullet, surface of the cartridge case and lubricants, which can be either of inorganic or organic nature.

Within this framework, the sampling carried out by the police when performing the physical examination of the crime scene or body for evidence of GSRs is a very important task. It must be taken into account that these residues may be found on the hands of the suspected shooter and in persons at close proximity or up to 30 feet from the barrel of the firearm. The deposition of GSR particles following initial firearm discharge is a primary transfer. A secondary transfer of these particles to other surfaces can occur by direct contact with the surfaces or persons on whom these particles have deposited, including handshaking and contact with clothing. Therefore, the movement of persons following the shooting, and even the activities carried out by forensic scientists at the crime scene, may alter GSRs distribution. Tertiary or even quaternary transfer is also possible, and law enforcement personnel may carry particles from prior shooting events.

The cartridge case, bullet, bullet coating, and metal jacket usually contain specific elements that can be detected during the analysis of GSRs. All cartridge cases are virtually made of brass (i.e., 70% Cu + 30% Zn) and a few of them have nickel coatings. Primer cartridge cases are of similar composition (Cu-Zn). Bullet cores are often made of lead and tin, with a very few having a ferrous alloy core. Bullet jackets are usually made of brass (90% Cu + 10% Zn), but some of them are of ferrous or aluminum alloys and some bullet coatings may also contain nickel.

Traditional methods for GSRs analysis are usually focused on the determination of inorganic compounds such as Pb, Ba and Sb. In many cases, inconclusive results can be obtained by these methods because, actually, most of these residues do not contain these compounds. Therefore, alternative methods are needed. In this sense, MEKC is a good option for the analysis of characteristic organic gunpowder components including NG, diphenylamine (DPA), ethyl centralite (EC) and others [148, 149].

Smokeless powder is the primary propellant in civilian and military ammunition and is the most common low explosive used to fabricate IEDs such as pipe bombs. Although this type of gunpowder is primarily composed of the propellant NC, it may also contain other substances, including the energetic material NG and the stabilizers DPA, N-nitrosodiphenylamine (NnDPA), and EC. The determination of the additive content has important forensic applications [149].

Considerable efforts have been made to develop analytical methods that are capable of identifying low levels of inorganic anions. The determination of nitrite and nitrate ions can be used as screening tools for investigating residues of firearm discharge due to the fact that these ions are the major inorganic components of GSRs. A rapid CZE method was developed by Erol et al. for the analysis of nitrate and nitrite in GSRs without using buffer additives [169]. In order to improve the detection limits, which was necessary especially for nitrite ion quantification, LVSS was employed as pre-concentration methodology.

However, the simultaneous determination of inorganic and organic compounds is highly desired. At this regard, a MEKC method has been developed for the simultaneous separation of several inorganic (i.e., Sb, Fe, Ba, Ca, Mg, Al, Ni, Zn, Pb and Cu) and organic (i.e., NG, resorcinol, 2,4-dinitrotoluene (DNT), 2,6-DNT, 2,3-DNT, dimethyl phtalate, diethyl phtalate, dibuthyl phtalate, methyl centralite, ethyl centralite and diphenylamine) components in GSRs [170]. Inorganic components were submitted to complexation with diaminocyclohexane tetraacetic acid (CDTA). CE is a cheaper, faster and more specific alternative than the common techniques usually employed for the analysis of real samples. MEKC was also used by MacCrehan et al. for the evaluation of residue contamination from organic additives in smokeless gunpowder [171]. The analysis of the residues showed only a trace of the previously fired power in the first shot and none in the subsequent shots.

On the other hand, numerous approaches have been used for the collection of the commonly determined primer-metal GSRs, usually from hands and clothing. Head hair has also been considered as a promising residue collection medium. A MEKC method has been developed for the analysis of organic GSRs collected from hair [172]. NG was detected by CE in the majority of the collection experiments and EC in some of them. In addition, a comparison between HPLC and MEKC has been performed for the analysis of organic components of smokeless gunpowder [173]. Both reverse-phase HPLC and MEKC techniques succeeded in resolving the tested standard mixtures of organic components of smokeless powders although the sequence of elution of the different analytes was slightly different.

Table 3 presents the CE-based methods recently published for the analysis of explosives and GSRs, showing that CE coupled with UV detection has been widely employed for the determination of inorganic and organic compounds.

Table 3. Applications of CE to the analysis of explosives and GSR

Analyte	Analytical technique	LODs	Type of sample	Reference
Cl^-, NO_2^-, NO_3^-, SO_4^{2-}, ClO_4^-, SCN^-, ClO_3^-, OCN^-, NH_4^+, K^+, Na^+, Ca^{2+}, Mg^{2+}	CZE-indirect UV	0.5 - 15 mg/L	Low explosives in post-blast residues	[151]
Cl^-, NO_3^-, SO_4^{2-}, SCN^-, Ca^{2+}, Fe^{2+}, Fe^{3+}	CZE-indirect UV	0.4 - 15 mg/L	Post-blast residues of flash, black and smokeless powders	[152]
Cl^-, ClO_4^-, NO_2^-, NO_3^-, SO_4^{2-}, NH_4^+, K^+	CZE-indirect UV	0.10 - 1.85 mg/L	Explosive residues	[153]
NO_2^-, NO_3^-, $S_2O_3^{2-}$, ClO_4^-, SCN^-, ClO_3^-, SO_4^{2-}, HCO_3^-, and HPO_4^{2-}, formate, K^+, NH_4^+, Na^+, Ca^{2+}, Mg^{2+} Ba^{2+}, Sr^{2+}, $CH_3NH_3^+$, Li^+, sucrose, fructose, glucose	CE-DAD	< 1.1 mg/L for ions	Explosive charge used in simulated suicide bombing attack	[155]
Cl^-, NO_2^-, NO_3^-, ClO_4^-, SCN^-, $S_2O_3^{2-}$, SO_4^{2-}, HCO_3^-, HPO_4^{2-}, N_3^-, ClO_3^-, formate, I^-, Br^-, F^-, OCN^-, acetate, oxalate, chlorite	CE-DAD	0.29 – 0.64 mg/L	Pre- and post-blast residues in soil, metal, plastic, paper, cotton and clothes	[154]
ClO_3^-, Cl^-, ClO_4^-	Chip-ITP-conductivity detector	0.8 - 1.75 mg/L	Flash power explosives	[156]
Cl^-, NO_3^-, N_3^-, ClO_4^-, ClO_3^-, SCN^-, HCO_3^-, SO_4^{2-}, F^-, HPO_4^{2-}	SI-CE-C^4D	23 - 79 µg/L	Soil extracts from detonated HMEs	[157]
15 anions + 12 cations	CE-indirect LED	0.11 - 2.30 mg/L	Post-blast residues from IEDs	[158]
NO_2^-, NO_3^-	CE-DAD	0.8 - 1.1 mg/L	Single-base gunpowder	[159]
15 anions + 11 cations	Portable CE-C^4D	27 - 240 µg/L	Pre- and post-blast residues	[160]
NH_4^+, K^+, $CH_3NH_3^+$, Na^+, Cl^-, NO_3^-, ClO_4^-	Chip-CE-C^4D	3.2 - 8.7 µM	Inorganic explosive residues	[161]

Table 3. (Continued)

Analyte	Analytical technique	LODs	Type of sample	Reference
NH_4^+, K^+, Na^+, Ca^{2+}, Mg^{2+}, Li^+, Cl^-, NO_3^-, ClO_4^-, ClO_3^-, SO_4^{2-}, F, CH_3SO^{3-}, HPO_4^{2-}	SI-CE-C^{4}D	5 - 61 µg/L	Environmental and industrial samples	[162]
25 organic and peroxide-based organic explosives	MEKC-DAD	5.5 - 45 mg/L	Pre- and post-blast residues	[164]
14 organic explosive compounds	MEKC and CEC IDLIF	1 - 10 mg/L	Pre-blast residues	[165]
14 nitroaromatic and nitramine explosives and explosive degradation products	MEKC-UV and CEC-UV	< 100 µg/L	Seawater samples	[166]
13 nitroaromatic and nitroamine explosives	CEC-amperometric detection	100 – 200 µg/L	Contaminated soils and ground water	[167]
14 organic compounds	CEC-UV and ToF-MS	1.0 - 3.2 mg/L	Commercial smokeless powders	[168]
NO_2^-, NO_3^-	LVSS-CE-UV	4.3 - 6.7 µM	Hands of shooters	[169]
Sb, dimethyl phtalate, Fe, diethyl phtalate, Ba, dibuthyl phtalate, Ca, methyl centralite, Mg, ethyl centralite Al, Ni, Zn, Pb, Cu, NG, resorcinol, 2,4-DNT, diphenylamine, 2,6-DNT, 2,3-DNT	MEKC-UV	0.11 - 10.94 mg/L	Real samples residues of pistol and revolvers on hands	[170]

ANALYSIS OF DYES IN INKS, FIBERS AND ARTWORK

From a forensic perspective, the analysis of dyes and pigments is of great relevance in the authentication of documents and artwork as well as in the analysis of fibers collected as trace evidence at the crime scene. Dyes are organic compounds and are used to produce inks and paints, and give color to textile materials. Pigments are particulate organic or inorganic solids and can be colored, colorless, or fluorescent compounds. Both dyes and pigments impart color by the selective absorption and reflection of certain visible wavelengths of light. Unlike pigments, dyes are soluble in the medium in which they are added. This section mainly deals with the analysis of dyes because they are more widely used than pigments in the production of inks and fibers.

Dyes consist of a wide range of compounds that, according to their applications methods, are usually classified in reactive, disperse, acid, basic, direct, and vat dyes [174]. Other classifications based on their chemical structure can also be found in the literature [175]. The color of inks, paints and textiles can be caused by a single dye or a combination of them. In many cases, different formulations comprising a dye mixture can lead to the same apparent color. For this reason, analytical methodologies are highly required to identify and distinguish dye formulations when carrying out certain types of forensic investigations.

The analysis of inks, which usually remains limited to the analysis of their dye components, is one of the main strategies to be followed during the investigation of questioned documents [176]. Indeed, ink analysis can result essential for verifying the authenticity of such documents [177, 178]. Questioned documents refer to those documents of doubtful authenticity or whose source is unknown, and may include letters, checks, wills, contracts, lottery tickets, passports and licenses. The authentication of documents represents the main issue dealt with by prosecution and justice bodies [9]. It must be carried out in cases of forgery, fraud, counterfeiting, contested wills or contracts, etc., but can also be required in incidents involving terrorist attacks, kidnappings or extortions where a letter may be left as evidence. Even in the case of suicide notes, they can be submitted to investigation in order to discard the covering-up of homicide. On the other hand, fibers from clothing and fabrics are one of the most important forms of trace evidence. In forensic investigations, textile fibers are excellent tools for linking suspects, victims and crimes scenes since they can prove the contact between individuals or between individuals and objects. Although the microscopical analysis of a fiber can reveal its major composition, the analysis of its colored components (i.e., dyes and pigments) provides additional and specific information, increasing the confidence on it as evidence in court [179]. Furthermore, and in addition to art conservation and restoration purposes [180], the analysis of dyes and pigments allows to detect art forgery.

The application of non-destructive techniques (e.g., microscopic and optical techniques, Raman spectroscopy) is always desirable for the analysis of inks in questioned documents, dyes in artwork or fibers. These objects constitute the evidence of a crime case, so their preservation must be ensured during forensic investigations. However, inks can be complex samples. They can also be contaminated by substances of the carrier (e.g., paper) or resulted from the aging process undergone by the document. Therefore, ink analysis sometimes requires the application of separation strategies (i.e., chromatographic techniques) in order to obtain a more comprehensive knowledge of their composition. Nevertheless, these are destructive methodologies. In this sense, the use of CE implies an advantage over traditional LC and GC methods. Only a few nanoliters of sample are needed when performing a CE separation, so the destruction of the examined material is minimal.

Considering the three types of crime cases mentioned above (i.e., investigation of questioned documents, fibers as trace evidences and art forgery), CE has found its main application in the analysis of inks from questioned documents. For this reason, this application is discussed in more detail by differentiating between methods intended for the analysis of writing inks, printer inks and stamp inks. On the other hand, CE has not been properly employed for detecting the falsification of artwork. However, in the field of conservation and restoration of cultural heritage, several CE methods have been developed for the identification of dyes and pigments in art pieces. The applicability of these methods can be easily extended to forensic purposes. CE has been applied to dye and ink analysis since 1991 [181]. Given the goal of this book to show the recent applications of CE, this section provides an overview of the CE-based methods applied to the analysis of dyes in inks, fibers and artwork, and developed during the early 21st century. For further information about previously reported methods, the reader is directed to other reviews found in the literature [143, 182].

Writing Inks

Despite the widespread use of printers, especially in official documents, the analysis of inks from hand-written documents such as checks is still carried out in criminalistics laboratories. In fact, more than 60% of the methods proposed for ink analysis are related to the analysis of writing inks [176]. There exists a wide range of writing ink formulations in the market. Pen inks are mainly classified in ballpoint inks (oil-based inks), fountain inks (water-based liquid inks), rollerball (water-based liquid or gel inks) and fiber-tip inks. Writing inks consist of several substances including coloring materials (dyes and pigments), solvents, resins, lubricants, driers, plasticizers, biocides, surfactants, waxes, etc. [183], being dyes the main ingredient analyzed when inks in questioned documents are investigated. The analysis of dyes by CE has mainly been accomplished for the discrimination of inks rather than for identification and quantification purposes. Nevertheless, several of the reported methods have shown satisfactory performance characteristics in terms of LODs and precision of migration times and peak areas [184, 185, 186].

In a forgery case, ink comparison is highly useful for establishing the relationship between two samples (i.e., an original document and a copy) as well as for detecting any alteration on the investigated document. As an example, the conviction of Martha Stewart represents a famous case in which the prosecution was partially based on ink-source-comparison evidence from a suspected stock worksheet alteration [187, 188]. Under this context, CE methods developed for ink analysis have been mainly used for the discrimination of inks by comparing their fingerprints without involving the identification of ink substances. In general, the majority of the reported CE methods are intended for the fingerprinting and discrimination of ballpoint pen inks as shown in Table 4. It follows the trend observed in this forensic field where more than 60% of the studies related to the analysis of writing inks in questioned documents are focused on ballpoint pen inks [176]. Ink formulations are usually under patent so the identification of their composition is very tedious, especially if MS is not used as detection mode. In this sense, when using CE coupled to DAD, the establishment of a database of UV spectra in addition to the electropherograms can be very useful for the identification and differentiation of inks [186].

Table 4. Applications of CE to the analysis of writing inks

Type of sample, sample taken and extraction solvent	Analytical technique and CE conditions	Reference
• Ballpoint pen ink • 1-10 dots (diameter of 1 mm) • EtOH	*NACE-LIF* BGE: 15 mM sodium borate and 835 mM acetic acid in EtOH (pH 5.4) Fused- silica capillary (75 µm ID, 50 cm total/30 cm effective length) Hydrodynamic injection: 5 cm for 15 s Voltage: 25 kV	[191]
• Blue pen inks (i.e., rollerball, ballpoint and marker pens) • 0.3 mg of the written paper collected by scratching • MeOH:water (1:1, v/v)	*CZE-DAD* BGE: 100 mM tris, 25% (v/v) MeCN (apparent pH 7.8) Fused- silica capillary (25 µm ID, 42 cm total/32 cm effective length) Hydrodynamic injection: 7 s at 1 psi Voltage: 15 kV; Temperature: 15°C	[186]
• Red ballpoint pen inks • 5-mm drawn line • MeOH	*NAMEKC-LIF* BGE: 3.5% (v/v) acetic acid, 100 mM SDS in formamide/water (9:1, v/v) Fused- silica capillary (75 µm ID, 38.5 cm total/29 cm effective length) Hydrodynamic injection: 5 s at 0.5 psi Voltage: 20 kV; Temperature: 25°C	[185]
• Blue ballpoint pen ink • Ink taken directly from the pen	*CZE-UV* BGE: 60 mM phosphate (pH 4.5) Fused-silica capillary (50 µm ID, 80 cm total/60 cm effective length) Voltage: 15 kV	[192]
• Black ballpoint pen inks • 1.0-cm drawn line • MeOH	*NACE-UV* BGE: 0.1% (v/v) acetic acid and 60 mM ammonium acetate in MeOH Fused- silica capillary (75 µm ID, 56 cm total/50 cm effective length) Hydrodynamic injection: 5 s at 0.5 psi Voltage: 30 kV; Temperature: 25°C	[184]

Type of sample, sample taken and extraction solvent	Analytical technique and CE conditions	Reference
• Fibre tip and ballpoint pen inks (black and blue inks) • Direct contact between the extraction solvent and the end of the pen • MeOH	*NACE-LED* BGE: 55 mM ammonium acetate and 600 mM acetic acid in MeOH Fused- silica capillary (50 μm ID, 48.5 cm total/40 cm effective length) Hydrostatic injection: 5 s at 100 mm height Voltage: 30 kV	[193]
• Black ballpoint pen inks • 1-5 dots (1 mm of diameter) /1-cm drawn line • MeOH	*CE-DAD* *Cationic dyes method:* BGE: 70/30 (v/v) acetate buffer (25 mM sodium acetate, 25 mM glacial acetic acid, 10 mM CTAB)/MeOH (pH 4.45, titrated with glacial acetic acid) Fused- silica capillary (75 μm ID, 40 cm total/30 cm effective length) Hydrodynamic injection: 5 s at 0.3 psi Voltage: 25 kV (reverse-polarity mode); Temperature: 30°C *Anionic dyes method:* BGE: 25 mM CHES and 5 mM β-cyclodextrin (pH 8.80, adjusted with 1.0 M NaOH) Fused- silica capillary (75 μm ID, 58 cm total/50 cm effective length) Hydrodynamic injection: 10 s at 25 mbar Voltage: 25 kV; Temperature: 30°C	[189]
• Fountain pen, gel pen and ballpoint pen inks • 2-4 dots (0.8 mm of diameter) • Pyridine:water (1:1, v/v)	*MEKC-DAD* BGE: 10.5 mM 3-aminopropanol, 5.25 mM HCl, 42 mM SDS, 0.35 mM Brij-35, 30% (v/v) MeCN Fused- silica capillary (50 μm ID, 75 cm total/57 cm effective length) Hydrodynamic injection: 6 s at 100 mbar Voltage: 30 kV	[194]

Researchers from the United States Federal Bureau of Investigation (FBI) have recently explored the potential of CE-DAD for ink identification, showing the importance of this technique for the investigation of questioned documents [188, 189]. The ultimate goal of these works was the establishment of libraries of black and blue ballpoint pen ink formulations for future forensic use. With the aim of characterizing ink samples, two CE methods were optimized for the separation of cationic dyes (general class of basic dyes) and the analysis of anionic dyes (general class of acidic dyes), respectively. Cationic dyes were separated in a mixture of acetate buffer and MeOH as BGE solution. The interactions between the positively charged analytes and the negatively charged silanol groups from capillary walls caused peak tailing, leading to a loss of peak resolution. Hexadecyltrimethylammonium bromide (CTAB) was added to the BGE solution for avoiding analyte-wall interactions. On the other hand, peak tailing was also observed when this method was applied to anionic dyes. The separation of anionic dyes was finally carried out in 2-(N-cyclohexylamino)ethanesulfonic acid (CHES) solution containing β-CDs, resulting in an improvement of peak efficiency. CDs are usually added to the BGE solution as an additive in order to separate chiral compounds or improve the separation between close-related compounds [190]. In the case of black ballpoint pen inks, more information was obtained when they were analyzed by the CE method for anionic dyes. This type of ink usually consists of anionic and neutral dyes, and only one single cationic dye (i.e., Crystal Violet/Methyl Violet) may be detected [189]. The CE method for cationic dyes resulted very effective for the differentiation of blue ballpoint pen inks because they are produced using mixtures of several cationic Victoria Blue dyes [188].

With the aim of analyzing acid and basic dyes under the same CE conditions, Calcerrada et al. have developed a CZE-DAD method for the determination of 9 dyes [186]. Although BGE conditions (apparent pH 7.8) were more favorable for the separation of acid dyes (i.e., Acid Blue, Acid Red and Acid Violet), electrophoretic peaks of 5 basic dyes (i.e., Basic Blue 9, Basic Violet 3, Basic Blue 11, Basic Blue 7, and Basic Blue 26) were partially resolved. The proposed CE method was finally applied to the differentiation of 34 blue pens, including 5 gel pens, 9 liquid pens and 20 ballpoint pens. Inks from different technologies as well as from different brand but the same technology were distinguished without any inconvenience. High discrimination power (100%) was achieved for the analysis of inks from the same brand but different model. Peak ratios were considered for the differentiation between inks from certain models. In addition, inks from the same brand and model but different batch were also differentiated. However, no differences were shown by inks from the same brand, model and batch.

One of the main problems faced by ink analysis is the low solubility in water of several types of inks. This supposes a disadvantage from two points of view: sample extraction and separation conditions. Oil-based inks such as ballpoint inks are effectively extracted from the carrier with organic solvents, usually using MeOH [184, 185, 193]. Nevertheless, this approach is not efficient for the extraction of water-based inks such as fountain inks. For this reason, it is preferable to apply a more generic method, involving the use of a mixture of MeOH:water (1:1, v/v) or pyridine:water (1:1, v/v) as extraction solvent [186, 194]. Regarding separation conditions, the hydrophobicity of certain ink components limits the applicability of CZE-based methods in ink analysis. To overcome this drawback, a surfactant (e.g., SDS or Brij-35) in combination with an organic solvent can be added to the separation medium for increasing sample solubility [194]. Under this condition, interactions between analytes and micelles, this means MEKC principles, must be considered [195]. Although MEKC is a plausible approach for ink analysis, especially for oil-based inks, NACE is often selected for this purpose [184, 191, 193].

Several organic solvents can be used as BGE solvent when performing NACE. MeCN or low molecular weight alcohols (i.e., MeOH, EtOH and propanol) can be used in NACE-UV or NACE-DAD methods. However, solvents exhibiting strong UV absorbance such as dimethylformamide (DMF) and dimethylsulfoxide (DMSO) are not suitable for UV-based methods [193, 196]. Zou et al. have carried out the separation of seven basic dyes (i.e., Methyl Violet, Crystal Violet, Tetramethyl-para-rosaniline, Victoria Blue B, Victoria Pure Blue, Rhodamine B and Rhodamine 6G) in MeOH containing acetic acid/ammonium acetate as BGE [184]. A total of 120 black ballpoint pen inks were analyzed by the proposed NACE-UV method and classified in 6 clusters according to the presence or absence of one dye. In addition to classical NACE methods, a non-aqueous micellar electrokinetic chromatography (NAMEKC) approach has recently been applied to the profiling of red ballpoint pen inks [185]. According to the authors, NAMEKC provides higher peak resolution than MEKC and NACE methods. In this sense, it must be taken into account that solvents traditionally used in NACE are not compatible with the formation of micelles and cannot be selected for NAMEKC separations. Formamide is an exception and surfactants such as SDS or CTAB are able to form micelles in it.

Despite UV has been widely used for ink fingerprinting [184, 186, 189, 194], it is a non-selective detection technique. Under this premise, other detection techniques must be considered for the differentiation of ink samples that present similar fingerprints when are analyzed by NACE-UV. As an alternative to UV detection, NACE has also shown to be compatible with LIF and light emitting diode (LED) absorbance detection [191, 193]. LIF is a more sensitive and selective technique than UV detection. However, the hyphenated CE-LIF systems, which are currently commercially available, possess an important limitation. Only one excitation wavelength can be selected during the analysis and it drastically reduces the number of peaks that can be detected. In order to enhance the detection capability of CE-LIF methods, Park et al. have recently developed a laboratory made CE-LIF system that allows the simultaneous detection of dyes and other ink substances exhibiting different excitation wavelengths [191]. On the other hand, there is a general lack of NACE-MS methods for the fingerprinting and identification of inks. Therefore, the development and implementation of this type of methods will provide a powerful tool for forensic examiners to distinguish between inks from questioned documents. MS allows the unequivocal identification of electrophoretic peaks. Consequently, it gives more confidence to the results obtained from the comparison of inks and increases the defensibility of the evidence [193].

In addition to the identification and discrimination of ink samples, ink dating is a challenge, but also a major goal, for forensic laboratories [176, 197]. It provides valuable information when analyzing questioned documents since it allows to detect any subsequent modification of the original document. It is important to take into account that ink composition may change over time due to storage conditions. In general, the use of CE for ink dating remains unexploited. In the last years, only one CE method has been developed for the dating of writing inks [192]. The photodegradation of Crystal Violet dye, which is commonly used in ink formulations, was monitored by CZE during 12 days. Methyl Violet 2B and Methyl Violet 6B were identified as its main degradation products. Future research for ink dating by CE should study the long-term degradation of inks due to natural aging processes.

Printer Inks

The forgery of documents by amateur and professional counterfeiters has increasingly grown since the widespread use of printers. The current printing technology produces forged documents of high quality, making difficult to detect any falsification or alteration. With the aim of preventing counterfeiting, several protection measures can be adopted such as the application of watermarks, holograms, invisible patterns, etc. However, these measures are not applied to all type of documents, so their authentication may require ink analysis. The analysis of printer inks, mainly focused on the comparison of inks, faces the same challenges as the analysis of writing inks (e.g., wide range of ink formulations, complex and unknown ink formulations, etc.). Nevertheless, specific analytical methods are required for the analysis of printer inks because they differ in composition from writing inks. CE methods for the analysis of black and color inkjet printing inks have been reported (Table 5). Until now, CE has not been use for the analysis of questioned documents involving other type of printing inks such as toner inks.

The investigation of inks from printed documents has been limited in comparison to the number of studies dealing with handwritten documents [176]. This is especially relevant if we only consider investigations carried out using CE as analytical technique. Considering the research works published since 2000, only one research group from Jagiellonian University (Kraków, Poland) has evaluated the potential and advantages of CE for the differentiation of printing inks.

Table 5. Applications of CE to the analysis of printer inks

Type of sample, sample taken and extraction solvent	Analytical technique and CE conditions	Reference
• Black inkjet printing inks • 10 dots (0.63 mm of diameter) • DMSO	*CZE-ESI-ToF-MS* BGE: 5 mM ammonium acetate, 0.01% (v/v) acetic acid, 25% (v/v) MeCN (apparent pH 5.41) Fused-silica capillary (50 μm ID, 100 cm total) Hydrodynamic injection: 7 s at 0.7 psi Voltage: 30 kV; Temperature: 25°C	[177]
• Black inkjet printing inks • 25 dots (ID 0.8 mm) • DMSO	*MEKC-DAD* BGE: 40 mM sodium borate, 20 mM SDS, 10% (v/v) MeCN (pH 9.5) Fused- silica capillary (75 μm ID, 60 cm total/50 cm effective length) Hydrodynamic injection: 6 s at 0.7 psi Voltage: 30 kV; Temperature: 25°C	[198]
• Color inkjet printing inks • 25 dots (0.8 mm of diameter) • BGE	*MEKC-DAD* BGE: 40 mM sodium borate, 20 mM SDS, 10% (v/v) MeCN (pH 9.56) Fused- silica capillary (75 μm ID, 60 cm total/50 cm effective length) Hydrodynamic injection: 6 s at 0.7 psi Voltage: 30 kV; Temperature: 25°C	[199, 200, 201]

Inkjet inks consist of a mixture of dyes and/or pigments (< 10%, w/w) and various additives (50-90%, w/w) in solvent, which are usually neutral substances. The analysis of other ink components in addition to dyes provides more complete fingerprints and leads to more accurate identification [177]. As a first approach, Szafarska et al. developed a MEKC-DAD method for the fingerprinting of color inkjet printing inks [199, 200, 201]. In order to extend the generated fingerprint database, this MEKC-DAD was further applied to the analysis of black inkjet inks [198]. MEKC was selected as CE mode with the aim of achieving the separation of neutral compounds contained in inkjet printing inks [200]. In comparison to CZE using borate buffer as BGE solution, MEKC improved the resolution and intensity of peaks as well as increased the number of peaks observed in the electrophoretic profiles. In terms of repeatability and reproducibility, high precision was achieved for migration times (RSDs < 3.3%), which is crucial for the comparison of ink fingerprints. Inks from five different manufacturers (i.e., Hewlett-Packard, Lexmark, Canon, Brother and Epson) were differentiated based on their electrophoretic fingerprint.

In a later work, the same authors analyzed more inks, confirming that color inks from different manufacturers and color inks with different composition but from the same supplier can be discriminated according to their electrophoretic profile [199]. Several questioned documents were also examined, including a document with a suspicious signature. The MEKC-DAD method demonstrated that the signature was generated by a printer although it was visually identified as a handwritten signature. Furthermore, MEKC-DAD has shown to provide higher discrimination power than Raman spectroscopy when comparing printouts produced with inkjet printing inks [201]. It must be considered that Raman spectroscopy is usually preferred over CE for the analysis of questioned documents because it is a non-destructive technique. However, CE requires low sample volume and documents can be analyzed with minimal damage.

It is well known that UV detection provides low sensitivity. Therefore, the quantity of sample required for CE analysis can be reduced even more if other detection modes such as MS are used. Only 10 paper dots (0.63 mm of diameter) were required to differentiate between black inkjet printing inks by CZE-MS [177], whereas their analysis by MEKC-DAD required of 25 paper dots (0.8 mm) [198].

To date, CE-MS has been scarcely evaluated for the analysis of writing or printing inks although MS provides more information than DAD or UV detection. This is highly relevant for the discrimination of black inkjet printing inks. MEKC-DAD has shown to be very ineffective for distinguishing between different black inks provided by the same manufacturer [198]. Under this context, a more complete fingerprint can be obtained by CE-MS, providing not only the migration time of each electrophoretic peak but also its related *m/z* ratio. In the case of fingerprints obtained by CE-UV or CE-DAD, peaks are mainly characterized by their migration time. Their related UV spectrum may not be sufficiently characteristic to distinguish between inks with confidence.

Kula et al. developed a CZE-ESI-ToF-MS method in order to distinguish between 13 black inkjet printing inks supplied by different manufacturers [177]. In the majority of inks investigated, it was observed a peak with a characteristic mass spectrum that was attributed to a surfactant or polymer (e.g., polyethylene glycol with different end groups). Inks from different suppliers were easily distinguished but, unlike MEKC-DAD, CZE-ESI-ToF-MS resulted in an efficient tool for the identification of black inkjet printing inks from the same brand but with different composition. However, the use of MEKC instead of CZE is advised in order to fully exploit the potential of CE-MS for ink discrimination. As observed by the authors, a wide range of ink substances migrated with the EOF under CZE conditions because they are neutral compounds. MEKC is widely employed for the separation of neutral molecules by CE but it has been avoided when using MS. Surfactants traditionally employed in MEKC separations (i.e., SDS) are non-volatile and tend to soil the mass spectrometer. However, recent MEKC-ESI-MS methods have demonstrated that alternative surfactants such as ammonium perfluorooctanoate (APFO) are compatible with MS [62, 202]. In addition, a partial filling approach can be followed for the successful combination of MEKC and MS [203].

Stamp Inks

In addition to writing and printing inks, forensics deal with the analysis of stamp inks. Many official and business documents are stamped as a guarantee of their authenticity, but stamps may be also falsified by forged stamp pads or using the current printing technology. Within this framework, CE methods already employed in the profiling of inkjet printing inks, including MEKC-DAD and CZE-MS methods [177, 200], have also shown to be applicable for the fingerprinting of stamp pad inks [204]. In comparison to printing inks which contain a wide range of substances, the differentiation between stamp inks is much more difficult because they often consist of only a single dye. Initially, 38 stamps inks (i.e., 9 green inks, 10 red inks, 9 violet inks and 10 blue inks) were characterized by their related electrophoretic profiles which were obtained by MEKC-DAD. Due to the variability of the migration times, each peak was defined by a relative migration time (ratio between its migration time and the migration time of peak identified as a compound coming from the paper). UV spectra of particular peaks were also considered for the differentiation between inks. Green and red stamps inks were easily distinguished from each other based on both electrophoretic and UV fingerprints. On the contrary, the analysis of violet inks and, especially, blue inks by MEKC-DAD did not provided a solid basis for claiming that they had a different origin. These inks were also submitted to CZE-MS analysis in order to obtain additional information for their identification. The majority of violet and blue inks were finally distinguished by their mass spectra. Additionally, the same composition was attributed to certain pairs of inks although they were acquired from different manufacturers. As an alternative to DAD and MS, LIF detection coupled to MEKC has also been evaluated for the discrimination of red stamp inks [205]. However, the proposed method demonstrated to be more effective for the differentiation of writing and printing inks.

Fibers

Textile fibers unquestionably represent one of the most common forms of trace evidence and are often the key piece of a criminal investigation. Recent studies show that they can be considered as solid evidence to link a suspect with the crime scene [206, 207]. Textile fibers are easily transferred, so fibers from victim's clothing or from the crime scene can be found on the clothing of the suspect. Moreover, fibers from suspect's clothing may be left as evidence at the crime scene [179, 208].

Textile fibers are classified in two main groups: (1) natural fibers, including vegetable (cellulose-based), animal (protein-based) and mineral fibers, and (2) man-made fibers, including fibers produced from natural fibers (natural polymers), synthetic fibers formed from non-renewable sources (synthetic polymers such as polyester) and inorganic fibers obtained from inorganic materials such as carbon or glass.

Table 6. Applications of CE to the analysis of dyes in fibers

Type of sample, sample taken and extraction solvent	Analytical technique and CE conditions	Reference
• Cotton fibers • 10-cm threads of fiber • Extraction solvent: (1) 35:65 (v/v) pyridine-water for direct dyes, (2) 1.5% (w/v) aqueous sodium hydroxide for reactive dyes, and (3) sodium dithionite solution for vat dyes	*CZE-DAD* BGE: 5 mM ammonium acetate, 40% (v/v) MeCN (pH 9.3) – 20 mM sodium dithionite (reducing agent) is added to the BGE solution for the analysis of vat dyes Fused-silica capillary (75 μm ID, 50 cm total/42 cm effective length) Hydrodynamic injection: 2 s at 50 mbar Voltage: 20 kV; Temperature: 25°C	[210]
• Acrylic fibers • One cm lengths of thread containing about 30-50 fibers • 88% (v/v) aqueous formic acid	*CZE-DAD and CZE-ESI-TOF-MS* BGE: 45 mM ammonium acetate (pH 4.7, adjusted with acetic acid), 60% (v/v) MeCN Fused-silica capillary (50 μm ID, 50 cm total/40 cm effective length) Hydrodynamic injection: 5 s at 1 psi Voltage: 20 kV; Temperature: 25°C	[211]
• Wool, polyamide, silk, viscose and cotton fibers • A single fiber of 2-15 mm length • Extraction solvent was selected according to the type of dye. As example, 4:3 (v/v) pyridine:water or ammonium:water was used for acid, direct dyes (see original paper for further information)	*ITP-MEKC-DAD* Leading ion: Cl^- (Na^+) (60 mM) Terminating ion: 3-[cyclohexylarnino]-1-propane sulfonic acid (CAPS)/$CAPS^-$ (15 mM) BGE: 15 mM 3-amino-1-propanol (3AP), 9.5 mM HCl (pH 9.7), SDS (60 mM), 30% (v/v) MeCN Fused- silica capillary (50 μm ID, 55 cm total/48 cm effective length) Hydrodynamic injection: 12 s at 700 mbar Voltage: 26 kV; Temperature: 44°C	[212]

The analysis of textile fibers gives a lot of information about their origin because they are physically and chemically distinguishable [209]. Microscopical examination is conducted in the early stages of fiber analysis. In addition to fiber nature and morphology, color is one of the main characteristics of fibers and reflects the dyes and pigments that were employed during their production process. Consequently, dye analysis is usually required to obtain greater scientific certainty about the matching of fiber pairs collected from different sources (i.e., suspect's clothing, victim's clothing, crime scene, etc.). Additionally, color is the only characteristic that can be reliably considered for the differentiation of certain type of fibers such as cotton [179]. Table 6 shows the latest methods reported for the analysis of dyes in fibers by CE.

Recently, Dockery et al. have evaluated a CZE-DAD method for the analysis of direct, reactive and vat dyes in cotton fibers [210]. The proposed CE method covers the determination of a large number of commercial dyes usually applied on cotton. Direct and reactive are soluble in water, so CZE involving an aqueous separation medium was suitable for their analysis. On the contrary, vat dyes are present in cotton as their oxidized *keto* form, being water-insoluble. In the abovementioned work, vat dyes were reduced to their *leuco* form with the aim of increasing their solubility in water. Under this condition, vat dyes showed to be soluble in the BGE solution (60:40 (v/v) water:MeCN) that contained sodium dithionite as reducing agent.

As occurs in the analysis of inks from questioned documents, the selectivity and sensitivity achieved by UV detection is limited when analyzing dyes from fibers by CE. Both performance characteristics can be improved by the use of alternative detection techniques such as MS [211], but sensitivity can also be enhanced by including a pre-concentration strategy in the CE method [213]. Xu et al. developed a sample-induced ITP-MEKC-DAD method for the determination of anionic dyes in wool, polyamide, silk, viscose and cotton fibers [212]. A large volume of sample was injected in the capillary (60% of capillary length), resulting in an enhancement in sensitivity between 190 and 640 times. The reproducibility of the measuring signal was high for single fibers from the same thread but that were extracted and analyzed separately. Consequently, this CE method makes possible the analysis of fibers in lengths encountered in forensic investigations (approximately between 2-20 mm).

Dyes and Pigments in Art Pieces and Ancient Graphic Documents

In the examination of an art piece, the analysis of dyes and pigments supports sample authentication. It reveals if the materials employed in its production were available to an artist at a particular time and place as well as if they were typically used by the author in his/her artistic work. At present, CE methods have not been developed for the detection of artwork forgery. By contrast, for conservation and restoration purposes, several methods have been reported for the identification of dyes and pigments in art pieces and ancient graphic documents (Table 7). Samples of the same or similar type (i.e., dyes present in paintings, drawings, etc.) are analyzed in both cases, so the application of these methods can be extended to criminal investigations related to artwork.

CZE-DAD has been successfully applied to the determination of five natural dyes (i.e., carmine, indigo, saffron, gamboge and *Rubia tinctoria* root) [214], as well as to identification of "sepia ink" [215], in maps dated between the 16th and the 19th century. In addition, MEKC-UV has been evaluated for the analysis of a broad range of red natural dyestuff (i.e., madder, cochineal, red sandalwood, brazilwood, lac dye and dragon's blood) [216]. These materials, also known as 'lake pigments', were identified by their coloring components namely alizarin, quinizarin and purpurin for madder, carminic acid for cochineal, santalin for red sandalwood, brazilin for brazilwood, laccaic acid for lac dye, and dracorhodin for dragon's blood. A non-attributed 17th century oil painting, property of the Diocesan Holy Art Museum of Bilbao (Spain), was submitted to analysis by the proposed MEKC-UV method. The presence of madder in the painting was confirmed by the detection of alizarin and the identification of low amounts of quinizarin and purpurin.

Furthermore, CZE-DAD and CZE-ESI-IT-MS have been recently proposed for the simultaneous screening, and further confirmation, of 23 cationic and anionic dyes used for autochromes [217]. The CZE-ESI-IT-MS method was applied to the analysis of a Filmcolor photograph, and four cationic dyes (i.e., Flexo Blue 810, Brillant Green, Rhodamine 6 G, and

Methylene Blue) and three anionic dyes (i.e., Rose Bengal, Diiodofluorescein, and Erythrosine B) were identified with certainty. In addition, other cationic dyes (i.e., Chrysoidine G and Crystal Violet) and anionic dyes (i.e., Patent Blue VF and Patent Blue V) were supposedly detected, but they were not unequivocally identified by MS^2 experiments due to low intensity of the electrophoretic peaks.

Table 7. Applications of CE to the analysis of dyes in art pieces and ancient documents

Type of sample, sample taken and extraction solvent	Analytical technique and CE conditions	Reference
• Photographs • 10 mm^2 of the artefact surface • Ethyl acetate	*CZE-DAD and CE-ESI-IT-MS* BGE: 50 mM ammonium acetate (pH 9 adjusted with 1 M NaOH), 15% (v/v) MeCN Fused-silica capillary (50 μm ID, 60 cm total/50 cm effective length) for CE-DAD. Fused-silica capillary (50 μm ID, 62 cm total length) for CE-MS. Hydrodynamic injection: 4 s at 1 psi Voltage: 20 kV; Temperature: 20°C	[217]
• Ancient documents • A brush imbibed with extraction solution was directly applied onto the document • 0.1 M SDS solution	*CZE-DAD* BGE: 20 mM sodium tetraborate (pH 9.2) Fused-silica capillary (50 μm ID, 64.5 cm total/56 cm effective length) Hydrodynamic injection: 4 s at 50 mbar Voltage: 25 kV; Temperature: 25°C	[215]
• Artwork (painting) • Paint chips (nanogram size samples) • Dichloromethane	*MEKC-UV* BGE: 20 mM borax (pH 9), 20 mM SDS, 10% (v/v) MeCN Fused-silica capillary (75 μm ID, 58 cm total/50 cm effective length) Hydrodynamic injection: 20 s Voltage: 20 kV; Temperature: 25°C	[216]
• Ancient documents • A brush imbibed with extraction solution was directly applied onto the document • 0.1 M SDS solution	*CZE-DAD* BGE: 40 mM sodium tetraborate (pH 9.25) Fused-silica capillary (50 μm ID, 64.5 cm total/56 cm effective length) Hydrodynamic injection: 13 s at 50 mbar Voltage: 30 kV; Temperature: 25°C	[214]

CONCLUSION

Nowadays, CE is a mature analytical technique that is well implemented in forensic laboratories. It is the principal tool for the analysis of forensic DNA, and has shown to be a good complementary approach to LC and GC for the identification and/or quantification of drugs of abuse, explosives, GSRs and inks from questioned documents. CE presents numerous advantages in comparison with traditional analytical techniques such as low solvent consumption and the requirement of small amounts of samples. This characteristic results especially attractive for forensic analysts because one of the main challenges of forensic science is that scientific evidences are not always found in great quantities. Nevertheless, the low concentration sensitivity provided by CE, particularly when coupled with UV-Vis detection, has limited its application in forensic analysis. Many efforts have been made at this regard and, as a result, several off-line and on-line pre-concentration strategies, including pre-concentration through sample treatments, have successfully been carry out for enhancing concentration sensitivity. Furthermore, the capabilities of this technique have been widely extended since its hyphenation with MS. In this sense, new CE methods based on UV-Vis, fluorescence and electrochemical detection are still developed for the analysis of forensic samples. However, CE-MS undoubtedly provides new insights for the determination of new drugs of abuse that are unknown but rapidly become very popular within drug consumers and for the unequivocal identification of trace evidences found at the crime scene (e.g., composition of fibers). In addition, great advances have been made in the miniaturization of CE systems (i.e., microchips). Under this context, forensic analysis may be performed in-situ at the crime scene, reducing the risk of evidence contamination as well as the time required for evidence analysis. As a result, forensic investigators can act more quickly to decision-making situations, limiting the reaction capacity of the suspects.

References

[1] Weinberger, R. (2000). *Practical Capillary Electrophoresis. Second Edition.* San Diego (CA, USA): Academic Press.

[2] Cruces-Blanco, C. (1998). *Electroforesis Capilar.* [*Capillary Electrophoresis*] Almería (Spain): Publication Services, University of Almería.

[3] Altria, K. D. (1996). *Capillary electrophoresis guidebook: principles, operation, and applications.* Totowa (NJ, USA): Humana Press Inc.

[4] Landers, J. P. (1996). *Handbook of Capillary Electrophoresis. Second Edition.* Florida (FL): CRC Press Inc.

[5] Manetto, G., Crivillente, F., and Tagliaro, F. (2000). Capillary electrophoresis: a new tool for forensic toxicologists. *Therapeutic Drug Monitoring*, 22:84-88.

[6] Heiger, D. N. (2000). *High-Performance Capillary Electrophoresis - An Introduction. Third Edition.* Germany: Agilent Technologies.

[7] Tagliaro, F., Turrina, S., and Smith, F. P. (1996). Capillary electrophoresis: Principles and applications in illicit drug analysis. *Forensic Science International* 77:211-229.

[8] Baldwin, R. P. (2000). Recent advances in electrochemical detection in capillary electrophoresis. *Electrophoresis* 21:4017-4028.

[9] Porpiglia, N., Giacomazzi, E., Gottardo, R., and Tagliaro, F. (2016). CE-MS in forensic sciences with focus on forensic toxicology. In *Capillary Electrophoresis – Mass Spectrometry (CE-MS): Principles and Applications. First Edition*, edited by Gerhardus de Jong, 217-291. Weinheim (Germany): Wiley-VCH Verlag GmbH & Co. KGaA.

[10] Hommerson, P., Khan, A. M., de Jong, G. J., and Somsen, G. W. (2011). Ionization techniques in capillary electrophoresis-mass spectrometry: Principles, design, and application. *Mass Spectrometry Reviews* 30:1096-1120.

[11] Mile, B. (2005). Chemistry in court. *Chromatographia* 62:3-9.

[12] Castillo-Peinado, L., and Luque de Castro, M. D. (2017). An overview on forensic analysis devoted to analytical chemists. *Talanta* 167:181-192.

[13] Favretto, D., Pascali, J. P., and Tagliaro, F. (2013). New challenges and innovation in forensic toxicology: focus on the "New Psychoactive Substances. *Journal of Chromatography A* 1287:84-95.

[14] Tagliaro, F., Pascali, J. P., and Lewis, S. W. (2013). Capillary Electrophoresis in Forensic Chemistry. In *Encyclopedia of Forensic Sciences, Second Edition*, edited by Jay A. Stegel and Pekka J. Saukko, 3:567-572. Waltham (MA, USA): Academic Press.

[15] Slepchenko, G. B., Gindullina, T. M., and Nekhoroshev, S.V. (2017). Capabilities of the electrochemical methods in the determination of narcotic and psychotropic drugs in forensic chemistry materials. *Journal of Analytical Chemistry* 72:703-709.

[16] Thormann, W., and Caslavska, J. (1998). Capillary electrophoresis in drug analysis. *Electrophoresis* 19:2691-2694.

[17] Tagliaro, F., Turrina, S., Pisi, P., Smith, F. P., and Marigo, M. (1998). Determination of illicit and/or abused drugs and compounds of forensic interest in biosamples by capillary electrophoretic/ electrokinetic methods." *Journal of Chromatography B* 713:27-49.

[18] Lurie, I. S. (1997). Application of micellar electrokinetic capillary chromatography to the analysis of illicit drug seizures. *Journal of Chromatography B* 780:265-284.

[19] Anastos, N., Barnett, N. W., and Lewis, S. W. (2005). Capillary electrophoresis for forensic drug analysis: A review. *Talanta* 67:269-279.

[20] Baciu, T., Botello, I., Borrull, F., Calull, M., and Aguilar, C. (2015). Capillary electrophoresis and related techniques in the determination of drugs of abuse and their metabolites. *TrAC Trends in Analytical Chemistry* 74:89-108.

[21] Tagliaro, F., Bortolotti, F. and Pascali, J. P. (2007). Current role of capillary electrophoretic/electrokinetic techniques in forensic toxicology. *Analytical and Bioanalytical Chemistry* 388:1359-1364.

[22] Cruces-Blanco, C. and García-Campaña, A. M. (2012). Capillary electrophoresis for the analysis of drugs of abuse in biological specimens of forensic interest. *TrAC Trends in Analytical Chemistry* 31:85-95.

[23] Kitagawa, F., and Otsuka, K. (2014). Recent applications of on-line sample preconcentration techniques in capillary electrophoresis. *Journal of Chromatography A* 1335:43-60.

[24] Rezaee, M., Assadi, Y., Hosseini, M. M., Aghaee, E., Ahmadi, F., and Berijani, S. (2006). Determination of organic compounds in water using dispersive liquid-liquid microextraction. *Journal of Chromatography A* 1116:1-9.

[25] Jain, R. and Singh, R. (2016). Applications of dispersive liquid-liquid micro-extraction in forensic toxicology. *TrAC Trends in Analytical Chemistry,* 75:227-237.

[26] Buszewski, B., and Szultka, M. (2012). Past, present, and future of solid phase extraction: A review. *Critical Reviews in Analytical Chemistry* 42:198-213.

[27] Pragst, F. (2007). Application of solid-phase microextraction in analytical toxicology. *Analytical and Bioanalytical Chemistry* 388:1393-1414.

[28] Takayama, N., Iio, R., Tanaka S., Chinaka, S. and Hayakawa, K. (2003). Analysis of methamphetamine and its metabolites in hair. *Biomedical Chromatography* 17:74-82.

[29] Boumba, V. A., Ziavrou, K. S., and Vougiouklakis, T. (2006). Hair as a biological indicator of drug use, drug abuse or chronic exposure to environmental toxicants. *International Journal of Toxicology* 25:143-163.

[30] Baciu, T., Borrull, F., Aguilar, C. and Calull, M. (2015). Recent trends in analytical methods and separation techniques for drugs of abuse in hair. *Analytica Chimica Acta,* 856:1-26.

[31] Han, E., Yang, H., Seol, I., Park, Y., Bongwoo, L., and Song, J. M. (2013). Segmental hair analysis and estimation of methamphetamine use pattern. *International Journal of Legal Medicine* 127:405-411.

[32] Barbosa, J., Faria, Carvalho, J., F., Pedro, M., Queirós, O., Moreira, R., and Dinis-Oliveira, R. J. (2013). Hair as an alternative matrix in bioanalysis. *Bioanalysis* 5:895-914.

[33] Tagliaro, F., Poiesi, C., Aiello, R., Dorizzi, R., Ghielmi, S., and Marigo, M. (1993). Capillary electrophoresis for the investigation of

illicit drugs in hair: determination of cocaine and morphine. *Journal of Chromatography* 638:303-309.

[34] Tagliaro, F., Valentini, R., Manetto, G., Crivellente, F., Carli, G., and Marigo, M. (2000). Hair analysis by using radioimmunoassay, high-performance liquid chromatography and capillary electrophoresis to investigate chronic exposure to heroin, cocaine and/or ecstasy in applicants for driving licences. *Forensic Science International* 107:121-128.

[35] Wernly, P., and Thormann, W. (1991). Analysis if illicit drugs in human urine by micellar electrokinetic capillary chromatography with on-column fast scanning polychrome absorption detection. *Analytical Chemistry* 63:2878-2882.

[36] Lombardo-Agüí, M., Cruces-Blanco, C., and García-Campaña, A. M. (2009). Capillary zone electrophoresis with diode-array detection for analysis of local anaesthetics and opium alkaloids in urine samples. *Journal of Chromatography B* 877:833-836.

[37] Alnajjar, A., Idris, A. M., Multzenberg, M., and McCord, B. (2007). Development of a capillary electrophoresis method for the screening of human urine for multiple drugs of abuse. *Journal of Chromatography B* 856:62-67.

[38] Botello, I., Borrullo, F., Aguilar, C., and Calull, M. (2012). Investigation of in-line solid-phase extraction capillary electrophoresis for the analysis of drugs of abuse and their metabolites in water samples. *Electrophoresis* 33:528-535.

[39] Xin, L., Rao, F., Min, L., Li-Ping, G., and Li, Y. (2013). Ionic liquid-based dispersive liquid-liquid microextraction coupled with capillary electrophoresis to determine drugs of abuse in urine. *Chinese Journal of Analytical Chemistry* 41: 1919-1922.

[40] Liotta, E., Gottardo, R., Bonizzato, L., Pascali, J. P., Bertaso, A., and Tagliaro, F. (2009). Rapid and direct determination of creatinine in urine using capillary zone electrophoresis. *Clinica Chimica Acta* 409:52-55.

[41] Breadmore, M. C., Wuethrich, A., Li, F., Phung, S. C., Kalsoom U., Cabot, J. M., Rokh, M. T., Shallan, A. I., Abdul Keyon, A. S., See H.

H., Dawod, M., and Quirino, J. P. (2017). Recent advances in enhancing the sensitivity of electrophoresis and electrochromatography in capillaries and microchips (2014-2016). *Electrophoresis* 38:33-59.

[42] Airado-Rodríguez, D., Cruces-Blanco, C., and García-Campaña, A. M. (2012). Dispersive liquid-liquid microextraction prior to field-amplified sample injection for the sensitive analysis of 3,4-methylenedioxymethamphetamine, phencyclidine and lysergic acid diethylamide by capillary electrophoresis in human urine. *Journal of Chromatography A* 1267:189-197.

[43] Quirino, J. P., and Terabe, S. (1998). Exceeding 5000-fold concentration of dilute analytes in micellar electrokinetic chromatography. *Science* 282:465-468.

[44] Lin, Y., Li, J., Ko, W., and Wu, S. (2006). Direct and sensitive analysis of methamphetamine, ketamine, morphine and codeine in human urine by cation-selective exhaustive injection and sweeping micellar electrokinetic chromatography. *Journal of Chromatography A* 1130:281-286.

[45] Cheng, H., Tsai, Y., Hsu, W., and Lin, Y. (2015). An on-line stacking capillary electrophoresis method for the analysis of Δ-9-tetrahydrocannabinol and its metabolites. *Journal of Chromatography A* 1426:226-332.

[46] Chiang, J., Hsiao, Y., Ko, W., and Wu, S. (2009). Analysis of multiple abused drugs and hypnotics in urine by sweeping CE. *Electrophoresis* 30:2583-2589.

[47] Ho, Y., Wang, C., Hsiao, Y., Ko, W., and Wu, S. (2013). Analysis of ten abused drugs in urine by large volume sample stacking-sweeping capillary electrophoresis with an experimental design strategy. *Journal of Chromatography A* 1295:136-141.

[48] Iwamuro, Y., Iio-Ishimaru, R., Chinaka, S., Takayama, N., Kodama, S., and Hayakawa, K. (2010). Reproducible chiral capillary electrophoresis of methamphetamine and its related compounds using a chemically modified capillary having diol groups. *Forensic Toxicology* 28:19-24.

[49] Alnajjar, A., and McCord, B. (2003). Determination of heroin metabolites in human urine using capillary zone electrophoresis with β-cyclodextrin and UV detection. *Journal Pharmaceutical and Biomedical Analysis* 33:463-473.

[50] Stege, P. W., Lapierre, A. V., Martínez, L. D., Messina, G. A., and Sombra, L. L. (2011). A combination of single-drop microextraction and open tubular capillary electrochromatography with carbon nanotubes as stationary phase for the determination of low concentration of illicit drugs in horse urine. *Talanta,* 86:278-283.

[51] Costa, J. L., Morrone, A. R., Ribeiro Resende, R., Aparecida da Matta Chasin, A., and Maggi Tavares, M. F. (2014). Development of a method for the analysis of drugs of abuse in vitreous humor by capillary electrophoresis with diode array detection (CE-DAD). *Journal of Chromatography B* 945-946:84-91.

[52] Baciu, T., Borrull, F., Aguilar, C., and Calull, M. (2016). Findings in the hair of drug abusers using pressurized liquid extraction and solid-phase extraction coupled in-line with capillary electrophoresis. *Journal of Pharmaceutical and Biomedical Analysis* 131:420-428.

[53] Lin, Y., Lee, M., Lee, R., Ko, W., and Wu, S. (2007). Hair analysis for methamphetamine, ketamine, morphine and codeine by cation-selective exhaustive injection and sweeping micellar electrokinetic chromatography. *Journal of Chromatography A* 1145:234-240.

[54] Webb, R., Doble, P., and Dawson, M. (2007). A rapid CZE method for the analysis of benzodiazepines in spiked beverages. *Electrophoresis* 28:3553-3565.

[55] Bishop, S. C., Lerch, M., and McCord, B. R. (2004). Micellar electrokinetic chromatographic screening method for common sexual assault drugs administrated in beverages. *Forensic Science International* 141:7-15.

[56] Iio, R., Chinaka, S., Tanaka, S., Takayama, N., and Hayakawa, K. (2013). Simultaneous chiral determination of methamphetamine and its metabolites in urine by capillary electrophoresis-mass spectrometry. *Analyst* 128:646-650.

[57] Botello, I., Borrull, F., Calull, M., Aguilar, C., Somsen, G. W., and de Jong, G. J. (2012). In-line solid-phase extraction-capillary electrophoresis coupled with mass spectrometry for determination of drugs of abuse in human urine. *Analytica and Bioanalytical Chemistry* 403:777-784.

[58] da Costa, J. L., Tonin, F. G., Zanolli, L. A., Aparecida da Matta Chasin, A., and Maggi Tavares, M. F. (2009). Simple method for determination of cocaine and main metabolites in urine by CE coupled to MS. *Electrophoresis* 30:2238-2244.

[59] Kohler, I., Schappler, J., and Rudaz, S. (2013). Highly sensitive capillary electrophoresis-mass spectrometry for rapid screening and accurate quantitation of drugs of abuse in urine. *Analytica Chimica Acta* 780:101-109.

[60] Kohler, I., Schappler, J., Sierro, T., and Rudaz, S. (2013). Dispersive liquid-liquid microextraction combined with capillary electrophoresis and time-of-flight mass spectrometry for urine analysis. *Journal of Pharmaceutical and Biomedical Analysis,* 73:82-89.

[61] Iwamuro, Y., Iio-Ishimaru, R., Chinaka, S., Takayama, N., and Hayakawa, K. (2012). Analysis of 11-nor-Δ^9-tetrahydrocannabinol-9-carboxylic acid and its glucuronide in urine by capillary electrophoresis/mass spectrometry. *Biomedical Chromatography* 26:1452-1456.

[62] Švidrnoch, M., Lněnickova, L., Válka, I., Ondra, P., and Maier, V. (2014). Utilization of micellar electrokinetic chromatography-tandem mass spectrometry employed volatile micellar phase in the analysis of cathinone designer drugs. *Journal of Chromatography A* 1356:258-265.

[63] Aturki, Z., D'Orazio, G., Rocco, A., Bortolotti, F., Gottardo, R., Tagliaro, F., and Fanali, S. (2010). CEC-ESI ion trap MS of multiple drugs of abuse. *Electrophoresis* 31:1256-1263.

[64] Gottardo, R., Polettini, A., Sorio, D., Pascali, J. P., Bortolotti, F., Liotta, E., and Tagliaro, F. (2008). Capillary zone electrophoresis (CZE) coupled to time-of-flight mass spectrometry (TOF-MS)

applied to the analysis of illicit and controlled drugs in blood. *Electrophoresis* 29:4078-4087.
[65] Boatto, G., Nieddu, M., Dessì, G., Manconi, P., and Cerri, R. (2007). Determination of four thiphenethylamine designer drugs (2C-T-series) in human plasma by capillary electrophoresis with mass spectrometry detection. *Journal of Chromatography A* 1159:198-202.
[66] Gottardo, R., Fanigliulo, A., Bortolotti, F., De Paoli, G., Pascali, J. P., and Tagliaro, F. (2007). Broad-spectrum toxicological analysis of hair based on capillary zone electrophoresis-time-of-flight mass spectrometry. *Journal of Chromatography A* 1159:190-197.
[67] Gottardo, R., Fanigliulo, A., Sorio, D., Liotta, E., Bortolotti, F., and Tagliaro, F. (2012). Monitoring compliance to therapy during addiction treatments by means of hair analysis for drugs and drug metabolites using capillary zone electrophoresis coupled to time-of-flight mass spectrometry. *Forensic Science International* 216:101-107.
[68] Gottardo, R., Bortolotti, F., De Paoli, G., Pascali, J. P., Mikšík, I., and Tagliaro, F. (2007). Hair analysis for illicit drugs by using capillary zone electrophoresis-electrospray ionization-ion trap mass spectrometry. *Journal of Chromatography A* 1159:185-189.
[69] Adcock, J. L., Barrow, C. J., Barnett, N. W., Conlan, X. A., Hogan, C. F., and Francis, P. S. (2011). "Chemiluminescence and electrochemiluminescence detection of controlled drugs." *Drug Testing Analysis* 3:145-160.
[70] Alnajjar, A., Butcher, J. A., and McCorde, B. (2004). Determination of multiple drugs of abuse in human urine using capillary electrophoresis with fluorescence detection. *Electrophoresis* 25:1592-1600.
[71] Frost, M., Köhlera, H., and Blaschke, G. (1997). Determination of LSD in blood by capillary electrophoresis with laser-induced fluorescence detection. *Journal of Chromatography B* 693:313-319.
[72] Zhang, L., Wang, R. Yu, Y. and Zhang, Y. (2007). Capillary electrophoresis with laser-induced fluorescence and pre-column

derivatization for the analysis of illicit drugs. *Journal of Chromatography B* 857:130-135.

[73] Ramseier, A., von Heeren, F., and Thormann, W. (1998). Analysis of fluorescein isothiocyanate derivatized amphetamine and analogs in human urine by capillary electrophoresis in chip-based and fused-silica capillary instrumentation. *Electrophoresis* 19:2967-2975.

[74] Elbshir, A. A., and Aboul-Enein, H. Y. (2012). Recent advances in applications of capillary electrophoresis with capacitively coupled contactless conductivity detection (CE-C^4D): an update. *Biomedical Chromatography* 26:990-1000.

[75] Epple, R., Blanes, L., Beavis, A., Roux, C., and Doble, P. (2010). Analysis of amphetamine-type substances by capillary zone electrophoresis using capacitively coupled contactless conductivity detection. *Electrophoresis* 31:2608-2613.

[76] Mantim, T., Nacapricha, D., Wilaira, P., and Hauser, P. C. (2012). Enantiomeric separation of some common controlled stimulants by capillary electrophoresis with contactless conductivity detection. *Electrophoresis* 33:388-394.

[77] Mazina, J., Saar-Reismaa, P., Kulp, M., Kaljurand, M., and Vaher, M. (2015). Determination of γ-hydroxybutiric acid in saliva by capillary electrophoresis coupled with contactless conductivity and indirect UV absorbance detectors. *Electrophoresis* 36:3042-3049.

[78] Gong, X. Y., Kubán, P., Scholer, A., and Hauser, P. C. (2008). Determination of γ-hydroxybutiric acid in clinical samples using capillary electrophoresis with contactless conductivity detection. *Journal of Chromatography A* 1213:100-104.

[79] Saar-Reismaa, P., Vaher, M., Kaljurand, M., Kulp, M., and Mazina-Šinkar, J. (2017). Simultaneous determination of γ-hydroxybutiric acid, ibotenic acid and psilocybin in saliva samples by capillary electrophoresis coupled with a contactless conductivity detector. *Analytical Methods* 9:3128-3133.

[80] Shewale, J. G., and Liu, R. H. (2014). *Forensic DNA Analysis: Current Practices and Emerging Technologies*. Boca Raton (FL): CRC Press Taylor and Francis Group.

[81] Jobling, M. A., and Tyler-Smith, C. (2017). Human Y-chromosome variation in the genome-sequencing era. *Nature Reviews Genetics* 18:485-497.

[82] Jeffreys, A. J., Wilson, V and Thein, S. L. (1985). Hypervariable 'minisatellite' regions in human DNA. *Nature* 314:67-73.

[83] Gill, P., Jeffreys, A. J., and Werret, David J. (1985). Forensic application of DNA 'fingerprints.' *Nature* 318:577-579.

[84] van Oorschot, R. A. H., Ballantyne, K. N., and Mitchell, R. J. (2010). Forensic trace DNA: a review. *Investigative Genetics* 1:14.

[85] Cobain, I. (2016). Killer breakthrough – the day DNA evidence first nailed a murderer. *The Guardian*, June 7. Accessed April 8, 2018. https://www.theguardian.com/uk-news/2016/jun/07/killer-dna-eviden ce-genetic-profiling-criminal-investigation.

[86] Butler, J. M. (2016). *Fundamentals of Forensic DNA Typing*. Amsterdam (Netherlands): Academic Press/Elsevier.

[87] Tagliaro, F., and Smith, F. P. (1996). Forensic capillary electrophoresis. *TrAC Trends in Analytical Chemistry* 15:513-525.

[88] Carey, L., and Mitnik, L. (2002). Trends in DNA forensic analysis. *Electrophoresis* 23:1386-1397.

[89] Tagliaro, F., and Bortolotti, F. (2006). Recent advances in the applications of CE to forensic sciences (2001-2004). *Electrophoresis* 27:231-243.

[90] Tagliaro, F., and Bortolotti, F. (2008). Recent advances in the applications of CE to forensic sciences (2005-2007). *Electrophoresis* 29:260-268.

[91] Tagliaro, F., Pascali, J., Fanigliulo, A., and Bortolotti, F. (2010). Recent advances in the applications of CE to forensic sciences (2007-2009). *Electrophoresis* 31:251-259.

[92] Pascali, J. P., Bortolotti, F., and Tagliaro, F. (2012). Recent advances in the applications of CE to forensic sciences (2009-2011). *Electrophoresis* 33:117-126.

[93] Thormann, W., Wey, A. B., Lurie, I. S., Gerber, H., Byland, C., Malik, N., Hochmeister, M., and Gehrig, C. (1999). Capillary

electrophoresis in clinical and forensic analysis: Recent advances and breakthrough to routine applications. *Electrophoresis* 20:3203-3236.

[94] Royal Society and the Royal Society of Edinburgh in conjunction with the Judicial College, the Judicial Institute, and the Judicial Studies Board for Northern Ireland. 2017. *Forensic DNA analysis: a primer for courts*. London (UK): The Royal Society.

[95] Yang, Y., Xie, B., and Yan, J. (2014). Application of next-generation sequencing technology in forensic science. *Genomics Proteomics Bioinformatics* 12:190-197.

[96] Alvarez-Cubero, M. J., Saiz, M., Martínez-García, B., Sayalero, S. M., Entrala, C., Lorente, J. A., and Martinez-Gonzalez, L. M. (2017). Next generation sequencing: an application in forensic sciences? *Annals of Human Biology* 44:581-592.

[97] Hares, D. R. (2015). Selection and implementation of expanded CODIS core loci in the United States. *Forensic Science International: Genetics* 17:33-34.

[98] Jobling, M. A., and Gill, P. (2004). Encoded evidence: DNA in forensic analysis. *Nature Reviews Genetics* 5:739:751.

[99] Buttler, J. M. (2012). Capillary electrophoresis: Principles and instrumentation. In *Advanced Topics in Forensic DNA Typing: Methodology*, edited by John M. Butler, 141-165. Amsterdam (Netherlands): Academic Press/Elsevier.

[100] Buttler, J. M., Buel, E., Crivellente, F., and McCord, B. R. (2004). Forensic DNA typing by capillary electrophoresis using the ABI Prism 310 and 3100 genetic analyzers for STR analysis. *Electrophoresis* 25:1397-1412.

[101] Moretti, T. R., Baumstark, A. L., Defenbaugh, D. A., Keys, K. M., Brown, A. L., and Budowle, B. (2001). Validation of STR typing by capillary electrophoresis. *Journal of Forensic Sciences* 46:661-676.

[102] Moretti, T. R., Baumstark, A. L., Defenbaugh, D. A., Keys, K. M., Smerick, J. B., and Budowle, B. (2001). Validation of short tandem repeats (STRs) for forensic usage: Performance testing of fluorescent multiplex STR systems and analysis of authentic and simulated forensic samples. *Journal of Forensic Sciences* 46:647-660.

[103] Cerri, N., Cisana, S., Cortellini, V., Verzeletti, A., Agostino, A., and De Ferrari, F. (2015). Validation of the AmpFLSTR® NGM select™ PCR amplification kit using ABI PRISM 310® Genetic Analyzer. *Forensic Science International: Genetics Supplement Series* 5:e350-e352.

[104] Adamowicz, M. S., Labonte, R., and Schienman, J. E. (2015). The potential of cosmetic applicators as a source of DNA for forensic analysis. *Journal of Forensic Sciences* 60:1001-1011.

[105] Bandhaya, A., and Panvisavas, N. (2008). Optimization of DNA recovery from toothbrushes. *Forensic Science International: Genetics Supplement Series* 1:9-10.

[106] Brettell, T. A., Butler, J. M., and Almirall, J. R. (2011). Forensic science. *Analytical Chemistry* 83:4539-4556.

[107] Sobrino, B., Brión, M., and Carracedo, A. (2005). SNPs in forensic genetics: a review on SNP typing methodologies. *Forensic Science International* 154:181-194.

[108] Sanchez, J. J., Phillips, C., Børsting, C., Balogh, K., Bogus, M., Fondevila, M., Harrison, C. D., Musgrave-Brown, E., Salas, A., Syndercombe-Court, D., Schneider, Peter M., Carracedo, A., and Morling, N. (2006). A multiplex assay with 52 single nucleotide polymorphisms for human identification. *Electrophoresis* 27:1713-1724.

[109] Prinz, M., Ishii, A., Coleman, A., Baum, H. J., and Shaler, R. C. (2001). Validation and casework application of a Y chromosome specific STR multiplex. *Forensic Science International* 120:177-188.

[110] Ballantyne, K. N., Keerl, V., Wollstein, A., Choi, Y., Zuniga, S. B., Ralf, A., Vermeulen, M., de Knijff, P., and Kayser, M. (2012). A new future of forensic Y-chromosome analysis: Rapidly mutating Y-STRs for differentiating male relatives and paternal lineages. *Forensic Science International: Genetics* 6:208-218.

[111] Melton, T. W., Holland, C. A. and Holland, M. M. (2012). Forensic mitochondrial DNA analysis: Current practice and future potential. *Forensic Science Review* 24:101-112.

[112] Foran, D. R., and Starrs, J. E. (2004). In Search of the Boston Strangler: Genetic evidence from the exhumation of Mary Sullivan. *Medicine, Science and the Law* 44:47-54.

[113] Sijen, T. (2015). Molecular approaches for forensic cell type identification: On mRNA, miRNA, DNA methylation and microbial markers. *Forensic Science International: Genetics* 18:21-32.

[114] Richard, L., Mara, L., Harper, K. A., Craig, R. L., Onorato, A. J., Robertson, J. M., and Donfack, J. (2012). Evaluation of mRNA marker specificity for the identification of five human body fluids by capillary electrophoresis. *Forensic Science International: Genetics* 6:452-460.

[115] Lindenbergh, A., de Pagter, M., Ramdayal, G., Visser, M., Zubakov, D., Kayser, M., and Sijen, T. (2012). A multiplex (m)RNA-profiling system for the forensic identification of body fluids and contact traces. *Forensic Science International: Genetics* 6:565-577.

[116] Durney, B. C., Crihfield, C. L., and Holland, L. A. (2015). Capillary electrophoresis applied to DNA: determining and harnessing sequence and structure to advance bioanalysis (2009-2014). *Analytical and Bioanalytical Chemistry* 407:6923-6938.

[117] Heller, C. (2001). Principles of DNA separation with capillary electrophoresis. *Electrophoresis* 22:629-643.

[118] Horvath, J., and Dolník, V. (2001). Polymer wall coatings for capillary electrophoresis. *Electrophoresis* 22:644-655.

[119] Barbier, V., and Viovy, Jean-Louis. (2003). Advanced polymers for DNA separation. *Current Opinion in Biotechnology* 14:51-57.

[120] Albarghouthi, M. N., and Barron, A. E. (2000). Polymeric matrices for DNA sequencing by capillary electrophoresis. *Electrophoresis* 21:4096-4111.

[121] Chiari, M., Cretich, M., and Consonni, R. (2002). Separation of DNA fragments in hydroxylated poly(dimethylacrylamide) copolymers. *Electrophoresis* 23:536-541.

[122] Barbier, V., Buchholz, B. A., Barron, A. E., and Viovy, J. (2002). Comb-like copolymers as self-coating, low-viscosity and high-

resolution matrices for DNA sequencing. *Electrophoresis* 23:1441-1449.

[123] Boulos, S., Cabrices, O., Blas, M., and McCord, B. R. (2008). Development of an entangled polymer solution for improved resolution in DNA typing by CE. *Electrophoresis* 29:4695-4703.

[124] Fredlake, C. P., Hert, D. G., Root, Brian E., and Barron, A. E. (2008). Polymer systems designed specifically for DNA sequencing by microchip electrophoresis: a comparison to commercially available materials. *Electrophoresis* 29:4652-4662.

[125] Castro, E. R., and Manz, A. (2015). Present state of microchip electrophoresis: State of the art and routine applications. *Journal of Chromatography A* 1382:66-85.

[126] Han, J., Gan, W., Zhuang, B., Sun, J., Zhao, L., Ye, J., Liu, Y., Li, C., and Liu, P. (2017). A fully integrated microchip system for automated forensic short tandem repeat analysis. *Analyst* 142:2004-2012.

[127] Karger, B. L., and Guttman, A. (2009). DNA sequencing by CE. *Electrophoresis* 30:S196-S202.

[128] Tsypryk, A., Gorbovitski, M., Kabotyanski, E. A., and Gorfinkel, V. (2006). Novel design of multicapillary arrays for high-throughput DNA sequencing. *Electrophoresis* 27:2869-2879.

[129] Hanning, A., Westberg, J., and Roeraade, J. (2000). A liquid core waveguide fluorescence detector for multicapillary electrophoresis applied to DNA sequencing in a 91-capillary array. *Electrophoresis* 21:3290-3304.

[130] Wan, F., Zhang, J., and Chu, B. (2009). Advances in electrophoretic techniques for DNA sequencing and oligonucleotide analysis. In *Advances in Chromatography, Volume 47*, edited by Eli Grushka, and Nelu Grinberg, 59-125. Boca Raton (FL): CRC Press Taylor and Francis Group.

[131] Han, J. P., Sun, J., Wang, L., Liu, P., Zhuang, B., Zhao, L., Liu, Y., and Li, C. X. (2017). The optimization of electrophoresis on a glass microfluidic chip and its application in forensic science. *Journal of Forensic Sciences* 62:1603-1612.

[132] Le Roux, D., Root, B. E., Reedy, C. R., Hickey, J. A., Scott, O. N., Bienvenue, J. M., Landers, J. P., Chassagne, L., and de Mazancourt, P. (2014). DNA analysis using an integrated microchip for multiplex PCR amplification and electrophoresis for reference samples. *Analytical Chemistry* 86:8192-8199.

[133] Yeung, S. H. I., Greenspoon, S. A., McGuckian, A., Crouse, C. A., Emrich, C. A., Ban, J., and Mathies, R. A. (2006). *Journal of Forensic Sciences* 51:740-747.

[134] Clayton, J. (2005). Go with the microflow. *Nature Methods* 2:621-627.

[135] Bruijns, B., van Asten, A., Tiggelaar, R., and Gardeniers, H. (2016). Microfluidic devices for forensic DNA analysis: A review. *Biosensors* 6:41.

[136] Horsman, K. M., Bienvenue, J. M., Blasier, K. R, and Landers, J. P. (2007). Forensic DNA analysis on microfluidic devices: A review. *Journal of Forensic Sciences* 52:784-799.

[137] Tetala, K. K. R., and Vijayalakshmi, M. A. (2016). A review on recent developments for biomolecule separation at analytical scale using microfluidic devices. *Analytica Chimica Acta* 906:7-21.

[138] Kim, Y. T., Heo, H. Y., Oh, S. H., Lee, S. H., Kim, D. H., and Seo, T. S. (2015). Microchip-based forensic short tandem repeat genotyping. *Electrophoresis* 36:1728-1737.

[139] Liu, P., Yeung, S. H. I., Crenshaw, K. A., Crouse, C. A., Scherer, J. R., and Mathies, R. A. (2008). Real-time forensic DNA analysis at a crime scene using a portable microchip analyzer. *Forensic Science International: Genetics* 2:301-309.

[140] Gibson-Daw, G., Albani, P., Gassmann, M., and McCord, B. (2017). Rapid microfluidic analysis of a Y-STR multiplex for screening of forensic samples. *Analytical and Bioanalytical Chemistry* 409:939-947.

[141] Kim, Y. T., Lee, D., Heo, H. Y., Sim, J. E., Woo, K. M., Kim, D. H., Im, S. G., and Seo, T. S. (2015). Total integrated slidable and valveless solid phase extraction-polymerase chain reaction-capillary

electrophoresis microdevice for mini Y chromosome short tandem repeat genotyping. *Biosensors and Bioelectronics* 78:489-496.

[142] Alonso, A., Albarran, C., Martín, P., García, P., Capilla, J., García, O., de la Rua, C., Izaguirre, N., Pereira, F., Pereira, L., Amorim, A., and Sancho, M. (2006). Usefulness of microchip electrophoresis for the analysis of mitochondrial DNA in forensic and ancient DNA studies. *Electrophoresis* 27:5101-5109.

[143] Cruces-Blanco, C., Gámiz-Gracia, L., and García-Campaña, A. M. (2007). Applications of capillary electrophoresis in forensic analytical chemistry. *TrAC Trends in Analytical Chemistry* 26:215-226.

[144] Calcerrada, M., González-Herráez, M., and García-Ruíz, C. (2016). Recent advances in capillary electrophoresis instrumentation for the analysis of explosives. *TrAC Trends in Analytical Chemistry* 75:75-85.

[145] Maitre, M., Kirkbride, K. P., Horder, M., Roux, C., and Beavis, A. (2017). Current perspectives in the interpretation of gunshot residues in forensic science: A review. *Forensic Science International* 270:1-11.

[146] Goudsmits, E., Sharples, G. P., and Birkett, J. W. (2016). Preliminary classification of characteristic organic gunshot residues compounds. *Science & Justice* 56:421-425.

[147] Meng, H., and Caddy, B. (1997). Gunshot residue analysis: A review. *Journal of Forensic Sciences* 42:553-570.

[148] Northrop, D. (2001). Gunshot residue analysis by micellar electrokinetic capillary electrophoresis: assessment for application to casework. Part II. *Journal of Forensic Sciences* 46:560-572.

[149] Northrop, D. (2001). Gunshot residue analysis by micellar electrokinetic capillary electrophoresis: assessment for application to casework. Part I. *Journal of Forensic Sciences* 46:549-559.

[150] Hargadon, K. A., and McCord, B. R. (1992). Explosive residue analysis by capillary electrophoresis and ion chromatography. *Journal of Chromatography A* 602(1–2): 241 –47.

[151] Hopper, K. G., LeClair, H., and McCord, B. R. (2005). A novel method for analysis of explosives residue by simultaneous detection

of anions and cations via capillary zone electrophoresis. *Talanta* 67:304-312.

[152] Ahmad, U. K., Azua Jamari, N. L., and Nagayah, S. A. L. (2012). Simultaneous determination of inorganic ions in post-blast residues using capillary electrophoresis. *Malaysian Journal of Analytical Sciences* 3:16-25.

[153] Feng, J., Guo, B., Lin, J., Xu, J., Zhou, H., Sun, Y., Liu, Y., Quan, Y., and Lu, X. (2008). Determination of inorganic anions in explosive residues by capillary zone electrophoresis. *Chinese Journal of Chromatography,* 26:667-671.

[154] Sarazin, C., Delaunay, N., Varenne, A., Vial, J., Costanza, C., Eudes, V., Mineta, J., and Gareil, P. (2010). Identification and determination of inorganic anions in real extracts frompre- and post-blast residues by capillary electrophoresis. *Journal of Chromatography A* 1217:6971-6978.

[155] Sarazin, C., Delaunay, N., Costanza, C., Eudes, V., and Gareil, P. (2013). On the use of capillary electrophoresis for the determination of inorganic anions and cations, and carbohydrates in residues collected after a simulated suicide bombing attack. *Talanta* 103:301-305.

[156] Prest, J. E., Beardah, M. S., Baldock, S. J., Doyle, S. P., Fielden, P. R., Goddard, N. J., and Treves Brown, B. J. (2008). Determination of chlorine containing species in explosive residues using chip-based isotachophoresis. *Journal of Chromatography A* 1195:157-163.

[157] Blanco, G. A., Nai, Y. H., Hilder, E. F., Shellie, R. A., Dicinoski, G. W., Haddad, P. R., and Breadmore, M. C. (2011). Identification of inorganic improvised explosive devices using sequential injection capillary electrophoresis and contactless conductivity detection. *Analytical Chemistry* 83:9068–9075.

[158] Hutchinson, J. P., Evenhuis, C. J., Johns, C., Kazarian, A. A., Breadmore, M. C., Macka, M., Hilder, E. F., Guijt, R. M., Dicinoski, G. M., and Haddad, P. R. (2007). Identification of inorganic improvised explosive devices by analysis of postblast residues using portable capillary electrophoresis instrumentation and indirect

photometric detection with a light-emitting diode. *Analytical Chemistry* 79:7005-7013.

[159] Alinat, E., Delaunay N., Costanza, C., Archer, X and Gareil, P. (2014). Determination of the nitrogen content of nitrocellulose by capillary electrophoresis after alkaline denitration. *Talanta* 125:174-180.

[160] Hutchinson, J. P., Johns, C., Breadmore, M. C., Hilder, E. F., Guijt, R. M., Lennard, C., Dicinoski, G., and Haddad, P. R. (2008). Identification of inorganic ions in post-blast explosive residues using portable CE instrumentation and capacitively coupled contactless conductivity detection. *Electrophoresis* 29:4593-4602.

[161] Wang, J., Pumera, M., Collins, G., Opekar, F., and Jelínekc, I. (2002). A chip-based capillary electrophoresis-contactless conductivity microsystem for fast measurements of low-explosive ionic components. *Analyst* 127:719-723.

[162] Gaudry, A. J., Guijt, R. M., Macka, M., Hutchinson, J. P., Johns, C., Hilder, E. F., Dicinoski, G. W., Nesterenko, P. N., Haddad, P. R., and Breadmore, M. C. (2013). On-line simultaneous and rapid separation of anions and cations from a simple sample using dual-capillary sequential injection-capillary electrophoresis. *Analytica Chimica Acta* 781:80- 87.

[163] Northrop, D. M., Martire, D. E., and MacCrehan, W. A. (1991). Separation and identification of organic gunshot and explosive constituents by micellar electrokinetic capillary electrophoresis. *Analytical Chemistry* 63:1038-1042.

[164] Johns, C., Hutchinson, J. P., Guijt, R. M., Hilder, E. F., Haddad, P. R., Macka, M., Nesterenko, P. N., Gaudry, A. J., Dicinoski, G. W., and Breadmore, M. C. (2015). Micellar electrokinetic chromatography of organic and peroxide-based explosives. *Analytica Chimica Acta* 876:91-97.

[165] Bailey, C. G., and Wallenborg, S. R. (2000). Indirect laser-induced fluorescence detection of explosive compounds using capillary electrochromatography and micellar electrokinetic chromatography. *Electrophoresis* 21:3081-3087.

[166] Giordano, B. C., Copper, C. L., and Collins, G. E. (2006). Micellar electrokinetic chromatography and capillary electrochromatography of nitroaromatic explosives in seawater. *Electrophoresis* 27:778-786.

[167] Hilmi, A., and Luong, J. H. T. (2000). In-line coupling capillary electrochromatography with amperometric detection for analysis of explosive compounds. *Electrophoresis* 21:1395-1404.

[168] de Perre, C., Corbin, I., Blas, M., and McCord, B. R. (2012). Separation and identification of smokeless gunpowder additives by capillary electrochromatography. *Journal of Chromatography A* 1267:259-265.

[169] Erol, Ö. Ö., Erdogán, B. Y., and Onar, A. N. (2017). Nitrate and nitrite determination in gunshot residue samples by capillary electrophoresis in acidic run buffer. *Journal of Forensic Sciences* 62:423-427.

[170] Bernal Morales, E., and Revilla Vázquez, A. L. (2004). Simultaneous determination of inorganic and organic gunshot residues by capillary electrophoresis. *Journal of Chromatography A* 1061:225-233.

[171] MacCrehan, W. A., Patierno, E. R., Duewer, D. L., and Reardon, M. L. (2011). Investigating the effect of changing ammunition on the composition of organic additives in gunshot residue (OGSR). *Journal of Forensic Sciences* 46:57-62.

[172] MacCrehan, W. A., Layman, M. J., and Secl, J. D. (2003). Hair combing to collect organic gunshot residues (OGSR). *Forensic Science International* 135:167-173.

[173] Cascio, O., Trettene, M., Bortolotti, F., Milana, G., and Tagliaro, F. (2004). Analysis of organic components of smokeless gunpowders: High-performance liquid chromatography vs. micellar electrokinetic capillary chromatography. *Electrophoresis* 25:1543-1547.

[174] Gürses, A., Açıkyıldız, M., Güneş, K. and Gürses, M. S. (2016). *Dyes and Pigments*. Switzerland: Springer International Publishing.

[175] Gregory, P. (1990). Classification of dyes by chemical structure. In *The Chemistry and Application of Dyes. Topics in Applied Chemistry,* edited by David R. Waring, and Geoffrey Hallas, 17-47. Boston (MA): Springer.

[176] Calcerrada, M., and García-Ruiz, C. (2015). Analysis of questioned documents: A review. *Analytica Chimica Acta* 853:143-165.

[177] Kula, A., Król, M., Wietecha-Posłuszny, R., Woźniakiewicz, M., and Kościelniak, P. (2014). Application of CE-MS to examination of black inkjet printing inks for forensic purposes. *Talanta* 128:92-101.

[178] LaPorte, G. M. (2016). Chemical analysis for the scientific examination of questioned documents. In *Forensic Chemistry: Fundamentals and Applications*, edited by Jay A. Siegel, 318-353. Oxford (United Kingdom): John Wiley & Sons, Ltd.

[179] Goodpaster, J. V., and Liszewski, E. A. (2009). Forensic analysis of dyed textile fibers. *Analytical and Bioanalytical Chemistry* 394:2009-2018.

[180] Degano, I., Ribechini, E., Modugno, F., and Colombini, M. P. (2009). Analytical methods for the characterization of organic dyes in artworks and in historical textiles. *Applied Spectroscopy Reviews* 44:363-410.

[181] Fanali, S., and Schudel, M. (1991). Some separations of black and red water-soluble fiber-tip pen inks by capillary zone electrophoresis and thin-layer chromatography. *Journal of Forensic Sciences* 36:1192-1197.

[182] Zlotnick, J. A., and Smith, F. P. (1999). Chromatographic and electrophoretic approaches in ink analysis. *Journal of Chromatography B* 733:265-272.

[183] Brunelle, R. I., and Crawford, K. R. (2003). *Advances in the Forensic Analysis and Dating of Writing Ink*. Springfield (IL): Charles C. Thomas Publisher, LTD.

[184] Zou, H., Wang, Z., Ye, N., Xu, R., Zhu, R., Wang, B., and Gu, X. (2008). NACE discrimination of black ballpoint pen inks. *Chromatographia* 67:483-486.

[185] Su, J., Xu, J., Zhong, S., Chen, L., Li, X., and Zou, H. (2014). Analysis of dye components in red ballpoint pen inks by non-aqueous micellar electrokinetic chromatography with LIF detection. *Chromatographia* 77:471-478.

[186] Calcerrada, M., González-Herráez, M., and Garcia-Ruiz, C. (2015). A microdestructive capillary electrophoresis method for the analysis of blue-pen-ink strokes on office paper. *Journal of Chromatography A* 1400:140-148.

[187] Hays, C. L. (2003). Prosecuting Martha Stewart: The Overview; Martha Stewart Indicted by U.S. On Obstruction. *The New York Times*, June 5. Accessed April 2, 2018. https://www.nytimes.com/2003/06/ 05/ business/ prosecuting-martha-stewart-overview-martha-stewart-indicted-us-obstruction.html.

[188] Brewer, J. D., Hagan, K. A., and Egan, J. M. (2005). Forensic analysis of blue ballpoint pen inks using capillary electrophoresis. *Forensic Science Communications*, 7. Accessed Mars 21 (2018). https:// archives.fbi.gov/archives/about-us/lab/forensic-science-communications/fsc/july2005/research/2005_07_research02.htm.

[189] Egan, J. M., Hagan, K. A., and Brewer, J. D. (2005). Forensic analysis of black ballpoint pen inks using capillary electrophoresis. *Forensic Science Communications,* 7. Accessed Mars 21 (2018). https:// archives.fbi.gov/archives/about-us/lab/forensic-science-communications/fsc/july2005/research/2005_07_research 01.htm.

[190] Escuder-Gilabert, L., Martín-Biosca, Y., Medina-Hernández, M. J., and Sagrado, S. (2014). Cyclodextrins in capillary electrophoresis: Recent developments and new trends. *Journal of Chromatography A* 1357:2-23.

[191] Park, M., Bahng, S., Woo, N., and Kang, S. H. (2016). Highly sensitive wavelength-dependent nonaqueous capillary electrophoresis for simultaneous screening of various synthetic organic dyes. *Talanta* 152: 236-243.

[192] Shih, C., Liu, J., Chen, B., and Lin, C. (2008). Separation of cristal violet dyes and its application to pen ink analysis using CZE and MEKC methods. *Journal of Separation Science* 31:893-897.

[193] Fakhari, A. R, Breadmore, M. C., Macka, M., and Haddad, P. R. (2006). Non-aqueous capillary electrophoresis with red light emitting

diode absorbance detection for the analysis of basic dyes. *Analytica Chimica Acta* 580:188-193.

[194] Mania, J., Bis, J., and Kościelniak, P. (2002). An evaluation of the application of capillary electrophoresis to forensic examination of inks. *Problems of Forensic Sciences* 51:71–86.

[195] Otsuka, K., and Terabe, S. (1996). Micellar electrokinetic chromatography. In *Capillary Electrophoresis Guidebook. Methods in Molecular Biology,* vol. 52, edited by Kevin D. Altria, 125-155. Totowa (NJ): Humana Press.

[196] Riekkola, M., Jussila, M, Porras, S. P., and Valkó, I, E. (2000). Non-aqueous capillary electrophoresis. *Journal of Chromatography A* 892:155-170.

[197] Ezcurra, M., Góngora, J. M. G., Maguregui, I, and Alonso, R. (2010). Analytical methods for dating modern writing instrument inks on paper. *Forensic Science International* 197:1-20.

[198] Król, M., Kula, A., Wietecha-Posłuszny, R., Woźniakiewicz, M., and Kościelniak, P. (2012). Examination of black inkjet printing inks by capillary electrophoresis. *Talanta* 96:236-242.

[199] Szafarska, M., Wietecha-Posłuszny, R., Woźniakiewicz, M., and Kościelniak, P. (2011). Application of capillary electrophoresis to examination of color inkjet printing inks for forensic purposes. *Forensic Science International* 212:78-85.

[200] Szafarska, M., Wietecha-Posłuszny, R., Woźniakiewicz, M., and Kościelniak, P. (2011). Examination of colour inkjet printing inks by capillary electrophoresis. *Talanta* 84:1234-1243.

[201] Król, M., Karoly, A., and Kościelniak, P. (2014). Raman spectroscopy and capillary electrophoresis applied to forensic colour inkjet printer inks analysis. *Forensic Science International* 242:142-149.

[202] Hernández-Mesa, M., Cruces-Blanco, C. and García-Campaña, A. M. (2017). Capillary electrophoresis-tandem mass spectrometry combined with molecularly imprinted solid phase extraction as useful tool for the monitoring of 5-nitroimidazoles and their metabolites in urine samples. *Talanta* 163:111-120.

[203] Muijselaar, P. G., Otsuka, K., and Terabe, S. (1998). On-line coupling of partial-filling micellar electrokinetic chromatography with mass spectrometry. *Journal of Chromatography A* 802:3-15.

[204] Król, M., Kula, A., and Kościelniak, P. (2013). Application of MECC-DAD and CZE-MS to examination of color stamp inks for forensic purposes. *Forensic Science International* 233:140-148.

[205] Król, M., Gondko, K., Kula, A., and Kościelniak, P. (2016). Analysis of red inks by micellar electrokinetic capillary chromatography with laser-induced fluorescence detection. *Electrophoresis* 37:372-380.

[206] Grieve, M., Biermann, T. W., and Schaub, K. (2005). The individuality of fibres used to provide forensic evidence – not all blue polyesters are the same. *Science & Justice* 45:13-28.

[207] Sauzier, G., Reichard, E., van Bronswijk, W., Lewis, S. W., and Goodpaster, J. W. (2016). Improving the confidence of "questioned versus known" fiber comparisons using microspectrophotometry and chemometrics. *Forensic Chemistry* 2:15-21.

[208] Deedrick, D. W. (2000). Hairs, fibers, crime, and evidence. Part 2: Fiber evidence. *Forensic Science Communications*, 2. Accessed April 1 (2018). https://archives.fbi.gov/archives/about-us/lab/forensic-science-communications/fsc/july2000/deedric3.htm.

[209] Robertson, J., Roux, C., and Wiggins, K. G. (2018). *Forensic Examination of Fibres, Third Edition*. Boca Raton (FL): CRC Press (Taylor & Francis Group, LLC).

[210] Dockery, C. R., Stefan, A. R., Nieuwland, A. A., Roberson, S. N., Baguley, B. M., Hendrix, J. E., and Morgan, S. L. (2009). Automated extraction of direct, reactive, and vat dyes from cellulosic fibers for forensic analysis by capillary electrophoresis. *Analytical and Bioanalytical Chemistry* 394:2095-2103.

[211] Stefan, A. R., Dockery, C. R., Baguley, B. M., Vann, B. C., Nieuwland, A. A., Hendrix, J. E., and Morgan, S. L. (2009). Microextraction, capillary electrophoresis, and mass spectrometry for forensic analysis of azo and methane basic dyes from acrylic fibers. *Analytical and Bioanalytical Chemistry* 394:2087-2094.

[212] Xu, X., Leijenhorst, H. A. L., Van Den Hoven, P., De Koeijer, J. A., and Logtenberg, H. (2001). Analysis of single textile fibres by sample-induced isotachophoresis – micellar electrokinetic capillary chromatography. *Science & Justice* 41:93-105.

[213] Breadmore, M. C., Tubaon, R. M., Shallan, A. I., Phung, S. C., Abdul Keyon, A. S., Gstoettenmayr, D., Prapatpong, P., Alhusban, A. A., Ranjbar, L., See, H. H., Dawod, M., and Quirino, J. P. (2015). Recent advances in enhancing the sensitivity of electrophoresis and electrochromatography in capillaries and microchips (2012–2014). *Electrophoresis* 36:36-61.

[214] López-Montes, A., Blanc García, R., Espejo, T, Huertas-Perez, J. F., Navalón, A., and Vílchez, J. L. (2007). Simultaneous identification of natural dyes in the collection of drawings and maps from The Royal Chancellery Archives in Granada (Spain) by CE. *Electrophoresis* 28:1243-1251.

[215] López-Montes, A., Blanc, R, Espejo, T. Navalón, A., and Vílchez, J. L. (2009). Characterization of sepia ink in ancient graphic documents by capillary electrophoresis. *Microchemical Journal* 93:121-126.

[216] Maguregui, M. I., Alonso, R. M, Barandiarán, M. J., Jiménez, R. M. T., and García, N. (2007). Micellar electrokinetic chromatography method for the determination of several natural red dyestuff and lake pigments used in art work. *Journal of Chromatograhy A*, 1154:429-436.

[217] López-Montes, A. M, Dupont, A. Desmazières, B., and Lavéndrine, B. (2013). "Identification of synthetic dyes in early colour photographs using capillary electrophoresis and electrospray ionization-mass spectrometry." *Talanta* 114:217-226.

In: Capillary Electrophoresis … ISBN: 978-1-53615-223-4
Editors: O. Núñez and X. Subirats

Chapter 4

ADVANCES IN THE CHIRAL ANALYSIS OF DRUGS BY CAPILLARY ELECTROPHORESIS

Elena Sánchez-López and María Luisa Marina*
Department of Analytical Chemistry,
Physical Chemistry and Chemical Engineering. University of Alcalá,
Alcalá de Henares (Madrid), Spain

ABSTRACT

The vast majority of drugs are chiral, thus, they can exist as two enantiomers. In most cases enantiomers display different pharmacological properties. Given the increasing demand on commercializing drugs in the enantiomerically pure format, there is the requirement to develop methodologies capable of resolving enantiomers. In this aspect, capillary electrophoresis has attracted much interest, thanks to its numerous advantages such as high efficiency, minimal sample and solvent consumption as well as its elevated flexibility in modifying the chiral environment, which increases the chances of achieving a chiral separation. The objective of this Chapter is to provide an insight on the recent contributions of capillary electrophoresis to the chiral analysis of drugs published from January 2016 to March 2018. All reported contributions

* Corresponding Author's Email: mluisa.marina@uah.es.

used UV detection being, in almost all cases, the electrokinetic chromatography mode using cyclodextrins, the strategy of choice.

Keywords: capillary electrophoresis, drug analysis, chiral separation, cyclodextrin, electrokinetic chromatography, enantiomer, pharmaceutical analysis.

INTRODUCTION

The individual determination of the enantiomers of racemic drugs has become an important issue in analytical chemistry and, more specifically, in drug analysis. This is due to the high number of drugs that are marketed as pure enantiomers nowadays and the need for assessing the enantiomeric purity of active principles (obtained both by asymmetric synthesis or preparative techniques) and pharmaceutical formulations and assuring the accomplishment of the related legal regulations. In fact, since one of the enantiomers of a racemic drug can have the desired biological activity while the other can be inactive, can have a different activity or can even be toxic, powerful analytical methodologies have to be developed enabling a reliable quality control of enantiomerically pure drugs.

Among the different separation techniques employed for chiral analysis, capillary electrophoresis (CE) has been recognized as a powerful technique to achieve enantiomeric separations due to its high efficiency, versatility and ease to change the chiral selector in the separation medium to obtain diverse enantioselectivities for a broad range of structurally different substances. Capillary electrokinetic chromatography (EKC), in which a chiral selector or a mixture of chiral selectors is added to the separation buffer, has been the most employed separation mode to carry out the chiral analysis of racemic drugs in the format of CE. The use of EKC and its different working modes such as micellar electrokinetic chromatography (MEKC) and microemulsion electrokinetic chromatography (MEEKC) together with different detection systems such as UV, laser-induced fluorescence, mass spectrometry and others, constitute very valuable tools to provide reliable analytical methods enabling the determination of the enantiomers of chiral drugs. A key point to mention is the so-called enantiomer migration order (EMO) since it is a crucial event as to have the enantiomeric impurity (distomer) to migrate before the active enantiomer (eutomer) to favour the determination of low amounts of the distomer. When compared to chromatographic techniques, CE offers relatively higher chances of reversing the EMO thanks to the fact that it is easier to modify the chiral media in CE [1].

Different chiral selectors are commercially available to be used in EKC. The most recent developments and applications (pharmaceutical and biomedical) achieved with chiral selectors in CE have been reviewed in a recent publication [2]. Native and derivatized cyclodextrins, antibiotics, polymeric surfactants, linear and branched oligo- and polysaccharides, cyclofructans, and others (chiral ionic liquids, metal-chiral acid complexes and DNA oligonucleotides) have been employed. Among them, the most widely applied chiral selectors were derivatized cyclodextrins. This fact was also confirmed by Q. Zhu et al. who reviewed the advances in the use of cyclodextrins as chiral selectors in EKC together with its fundamentals and applications in the pharmaceutical, biomedical and environmental fields [3] including the development of new CDs (especially positively charged) to change the enantioselectivity.

Related to the analysis of pharmaceuticals, the advances on the use of cyclodextrins in the chiral analysis of drugs by CE was reviewed showing the potential of these chiral selectors and the importance of its proper selection to achieve enantiomeric separations with application to the quantitative analysis of pharmaceutical formulations and their enantiomeric purity, the degradation studies of drugs, the chiral analysis of drugs in biological samples, the study of drug metabolism or the criminalistics and forensic investigations [4]. Also, the separation and analysis of drug-related substances, the use of chiral CE and chiral CE-MS, were reviewed [5, 6], showing the separation power of CE techniques that may be considered as the option of choice for chiral separations, not only due to its high efficiency and easy change of the chiral selector but also due to the reduced cost compared to HPLC [5]. Moreover, despite the limited sensitivity usually obtained by CE with UV detection, chiral impurities could often be detected in the 0.1% range [6].

In this chapter, the most recent advances achieved in the chiral analysis of drugs by CE are described. Articles published in the period of time from January 2016 to March 2018 are reviewed. The results obtained when using different chiral selectors are discussed and the most important characteristics of the developed methods are presented. Also, the most relevant applications reported on the chiral determination of drugs in pharmaceutical formulations and biological samples will be described.

Chiral Separation of Drugs by Capillary Electrophoresis

Determination of drugs is an important topic in separation sciences. Given the increasing interest in commercialization of pharmaceuticals as single enantiomers, the use of CE in the determination of the enantiomeric impurity is empowered thanks to its great advantages derived from its feasibility in the chiral analysis and its high efficiency, among others.

In the period of time covered in this Chapter many contributions regarding the chiral separation of drugs have been reported. Here, only those concerning the determination of chiral drugs in real samples will be discussed. Among the different samples to be analysed, we can sort them out by the analysis of pharmaceutical and biological samples, detailed in Table 1 and Table 2, respectively. As can be seen in both Tables, research works on drug analysis published in last years were carried out using solely UV detection, mainly at a detection wavelength of 200 nm. Regarding the CE mode applied, EKC has been the one most used, being in all cases aqueous CE used, except in two cases [18, 19] where non-aqueous CE (NACE) was the chosen mode.

The common strategy to develop chiral methods in CE has been focused on optimizing one variable at a time. Nonetheless, in last years there is an increasing interest in using design of experiments approaches in order to evaluate in a better way how different conditions are involved in improving enantioresolution. For instance, researchers have been using the so-called quality by design (QbD) principles [7-10]. QbD might be applied to establish a design space in which the analysis parameters meet a preselected quality attributes.

Determination of Drugs in Pharmaceutical Samples

Pharmaceuticals can vary from the bulk drug up to the finished product, such as tablets, capsules or injection solutions, among others. Sample treatment of pharmaceuticals is commonly a simple procedure which includes grinding the pharmaceutical and homogenizing the resulting powder to then bring it into solution using any suitable solvents such as water, DMSO or solvent mixtures. If pharmaceuticals are already in liquid state the common procedure is dilution. In pharmaceutical analysis, it is crucial to enable to determine the lowest level of enantiomeric impurity (distomer) as possible in presence of large amounts of the desired or active enantiomer (eutomer), also known as relative limit of detection (RLOD). On this matter, the International Conference on Harmonization (ICH) states that a method must be capable of detecting up to 0.1% of the distomer [33]. As can be seen in Table 1, most of contributions herein included are in agreement with the ICH guidelines, being the lowest RLOD 0.02% for different drugs [7, 8, 13]. Different drugs have been determined in pharmaceutical formulations such as antitussive, proton pump inhibitors, antibacterial, anticancer and β-blockers, among others. Details of these contributions are summarized below.

Levomepromazine is a known neuroleptic drug used in the treatment of schizophrenia and terminal pain and represents the R-enantiomeric configuration due to its higher affinity towards dopaminergic and serotoninergic receptors when compared to the S-enantiomer [34]. In a work by Niedermeier and Scriba [7], using QbD principles, development of a CE method was carried out to control, not only the enantiomeric purity of levomepromazine, but also to track its possible oxidation to levopromazine sulfoxide, an important impurity. This EKC-UV method based on the use of HP-γ-CD as chiral selector, was applied to the analysis of a reference standard of the European Pharmacopoeia and a commercial injection solution. On the one hand, high levels of dextromepromazine (S-enantiomeric impurity) were found only in the reference standard whereas that in the injected pharmaceutical solution it was below the LOQ. On the other hand, levomepromazine sulfoxide was below the LOD and below the LOQ for the reference standard and the injection solution, respectively. Authors pointed out the need to perform quick analysis since oxidation of levomepromazine quickly occurs when samples are left in the sampling tray, despite the use of antioxidants and low temperatures [7]. Another pharmaceutical studied in last years was dextromethorphan, a drug which possesses antitussive effects whereas that its enantiomeric counterpart, levomethorphan, is an opioid drug with analgesic effects. Once again, the QbD strategy was used to properly optimize the chiral method in CE [8]. Optimized method comprised the use of a dual cyclodextrin setup based on a combination of neutral M-α-CD and negatively charged S-β-CD. Capsules containing dextromethorphan were analysed and, in all cases, levomethorphan was not detected (it was below the 0.02% level) [8]. Additionally, QbD principles were also used in two contributions dealing the enantioseparation of ambrisentan [9, 10]. Ambrisentan is a drug used in the treatment of pulmonary arterial hypertension and chronic thromboembolic pulmonary hypertension. This drug is marketed as S-enantiomer due to its higher activity so that the distomer (R-ambrisentan) need to be controlled. In the first paper, authors developed a MEKC method based on the use of γ-CD and SDS [9]. Analyzed tablets did not display presence of enantiomeric impurity. In the second paper, authors could detect relative concentrations of 0.1% for the enantiomer impurity of ambrisentan (R-enantiomer) [10]. Researchers applied the method to the analysis of a chemical supplier sample in order to evaluate the enantiomeric purity of S-ambrisentan, in which values of about 0.15% of the enantiomeric impurity were found. Differences in the LOD were found for these two works, these being 1.5 [9] and 0.2 μg/mL [10], respectively.

Table 1. Contributions included in this Chapter regarding chiral drug analysis in pharmaceutical samples

Drug	Sample	BGE	CE mode and detection	LOD (RLOD)	Ref
Dextromepromazine and levomepromazine sulfoxide	Injection solution and a reference substance of the European Pharmacopoeia	3.6 mg/mL HP-γ-CD in 100 mM citric acid buffer (pH 2.85)	EKC-UV (253 nm)	0.03% (dextromepromazine) 0.02% (levomepromazine sulfoxide)	[7]
Dextromethorphan	Capsules	16 mg/mL S-β-CD and 14 mg/mL M-α-CD in 30 mM sodium phosphate buffer (pH 6.5)	EKC-UV (200 nm)	0.3 µg/mL (0.02%)	[8]
Ambrisentan	Tablets	50 mM γ-CD and 100 mM SDS in 100 mM borate buffer (pH 9.20)	MEKC-UV (200 nm)	1.5 µg/mL	[9]
Ambrisentan	A chemical supplier sample	30 mM γ-CD in 50 mM sodium acetate buffer (pH 4.0)	EKC-UV (200 nm)	0.2 µg/mL	[10]
Vildagliptin	Tablets	20 mM SBE-β-CD in 75 mM Tris-acetic acid (pH 4.75)	EKC-UV (200 nm)	2.5 µg/mL (0.14%)	[11]
Sitafloxacin	Three batches of pharmaceutical samples	15 mM D-Phe, 20 mM $CuSO_4$, 20 mM γ-CD in 15 mM dipotassium hydrogenophosphate (pH 4.5)	EKC-UV (297 nm)	3 - 4 µg/mL for sitafloxacin and its three impurities	[12]
Montelukast enantiomeric and diastereoisomeric forms	Bulk product, chewable tablets and oral granules (pouches)	10 mM SBE-β-CD, 10 mM TM-β-CD and 10 mM SDS in 20 mM borate (pH 9.0)	MEKC-UV (254 nm)	0.30 µg/mL (0.02%)	[13]
Dexlansoprazole	Capsules	30 mM SBE-β-CD and 20 mM methyl-β-CD in 90 mM phosphate buffer (pH 6.0)	EKC-UV (280 nm)	RLOD 0.07%	[14]
Colchicine	Tablets	7 mM succinyl-γ-CD in 50 mM borate buffer (pH 9.0)	EKC-UV (243 nm)	RLOD 0.1%	[15]

Table 1. (Continued)

Drug	Sample	BGE	CE mode and detection	LOD (RLOD)	Ref
Cinacalcet	Tablets	0.5% HP-γ-CD in 150 mM sodium phosphate (pH 2.5) containing 20% methanol	EKC-UV (214 nm)	0.14 μg/mL	[16]
Praziquantel	Tablets	15 mM S-β-CD in 50 mM phosphate buffer (pH 2.0)	EKC-UV (200 nm)	0.75 μg/mL	[17]
Clenbuterol	Oral solution	80 mM diacetone-D-mannitol, 100 mM boric acid and 50.4 mM triethylamine in methanol	NACE-EKC-UV (214 nm)	0.25 μg/mL	[18]
Propranolol	Tablets	40 mM D-(+)-xylose, 100 mM boric acid, and 78.9 mM triethylamine in methanol	NACE-EKC-UV (214 nm)	0.25 μg/mL (R-propranolol) and 0.50 μg/mL (S-propranolol)	[19]
Naproxene	Bulk sample	2% SBE-β-CD(w/v) + 1.2% (w/v) nanoliposome in 20 mM PBS (pH 5.2)	LEKC-UV (235 nm)	4.5 μg/mL	[20]

Abbreviations: EKC: electrokinetic chromatography; HP-γ-CD: hydroxypropyl-γ-CD; LEKC: liposome electrokinetic chromatography; LOD: limit of detection; M-α-CD: methyl-α-CD; MEKC: micellar electrokinetic chromatography; NACE: non-aqueous capillary electrophoresis; PBS: phosphate buffer solution; Phe: phenylalanine; RLOD: relative limit of detection; SBE-β-CD: sulfobutyl ether-β-CD; SDS: sodium dodecyl sulfate; S-β-CD: sulfated-β-CD; TM-β-CD: trimethyl-β-CD.

Table 2. Contributions included in this Chapter regarding chiral drug analysis in biological samples

Drug	Sample	BGE	CE mode and detection	LOD	Ref
Chlorpheniramine	Rat plasma	20 mg/mL S-β-CD in 30 mM Tris (pH 3.5) with 5% (v/v) methanol	EKC-UV (210 nm)	0.075 μg/mL	[21]
Tranylcypromine	Spiked human urine	7% (w/v) HP-α-CD in 100 mM phosphate buffer (pH 2.0)	EKC-UV (214 nm)	6.06 ng/mL	[22]
Tetrabenazine	Human liver microsomes	1% CM-β-CD in 80 mM tris-phosphate buffer (pH 2.5)	EKC-UV (240 nm)	Not stated	[23]
Pyroglutamic acid derivative	Raw material and rat plasma	2.5% (w/v) HS-β-CD in 50 mM phosphate buffer (pH 2.5)	EKC-UV (260 nm)	0.15 μg/mL	[24]
Aminoalkanol derivatives	Aqueous medium and serum	5% γ-CD in 25 mM phosphate buffer (pH 2.5)	EKC-UV (200 nm)	141 ng/mL	[25]
Ketamine, norketamine, 6-hydroxynorketamine and 5,6-dehydronorketamine	Equine liver microsomes	0.66% HS-γ-CD in 100 mM phosphate buffer (pH 3.0)	EKC-UV (195 nm)	Not stated	[26]
Ketamine and metabolites	Plasma and serum	0.66% HS-γ-CD in phosphate buffer (pH 3.0)	EKC-UV (200 nm)	~ ppb range	[27]
Ketamine and metabolites	Dog plasma	0.66% HS-γ-CD in 100 mM disodium hydrogenphosphate buffer (pH 3.0)	EKC-UV (200 nm)	~ ppb range	[28]
Ketamine and norketamine	Human hair	0.1% HS-γ-CD in 15 mM tris phosphate (pH 2.5)	EKC-UV (200 nm)	0.08 ng/mg	[29]
Methadone	Exhaled breath condensate, serum, and urine	0.8% CM-β-CD in 150 mM phosphoric acid-triethylamine (pH 2.5) containing 30% methanol	EKC-UV (200 nm)	0.15μg/mL (lower LOQ)	[30]
Methadone	Human hair	11 mM α-CD in 80 mM sodium phosphate (pH 2.5)	EKC-UV (200 nm)	0.13 ng/mg	[31]
Cathinone derivatives	Spiked human hair	12 mg/mL β-CD in 80 mM disodium phosphate (pH 2.5)	EKC-UV (200 nm)	0.02 - 0.10 ng/mg	[32]

Abbreviations: CM-β-CD: carboxymethyl-β-CD; EKC: electrokinetic chromatography; HP-α-CD: hydroxypropyl-α-CD; HS-β-CD: highly sulfated-β-CD; HS-γ-CD: highly sulfated-γ-CD; S-β-CD: sulfated-β-CD.

Vildagliptin is used in the treatment of type II diabetes. Since the S-enantiomer is the only enantiomer used in clinics, R-vildagliptin is considered an impurity. Once selected sulfobutyl-ether-α-CD as the most appropriate chiral selector for vildagliptin enantioseparation, researchers optimized other parameters, following a factorial analysis study [11]. Detection of up to 0.14% of the enantiomeric impurity was possible and analysis of enantiopure tablets did not display detectable amounts of the distomer.

Even though the common scenario is that drugs possesses only one chiral center, in some reports, chiral separation of drugs having two or more asymmetric centers has also been reported. For obvious reasons, in these cases, separation is a more complicated task. Antibacterial sitafloxacin possesses 3 chiral centers, meaning that eight possible stereoisomers might occur. Due to the fact that some stereoisomers have demonstrated to be cytotoxic it is important to control its stereoisomeric composition. Meng et al. studied the separation of 4 of the stereoisomers of sitafloxacin [12]. On the one hand, negatively charged HS-γ-CD enabled baseline chiral separation but the enantiomer migration order was not the desired one (see Figure 1A). On the other hand, a dual system based on γ-CD and a Cu^{2+}-D-phenylalanine complex enabled enantioseparation achieving the optimum EMO (Figure 1D). Interestingly, when using these two chiral selectors individually (see Figure 1B and Figure 1C) chiral separation was not achieved. This states the utmost necessity to use these two selectors combined [12]. Then, authors used response surface methodology with a face centered central composite design to continue optimizing the chiral method. Validation was further carried out, obtaining LODs from 3 to 4 μg/mL for sitafloxacin and its three impurities. When analyzing three batches of real samples containing sitafloxacin, stereoisomeric impurities were found below the LOD in all cases [12]. Another contribution based on separation of drug containing more than one chiral center was also reported. Montelukast is a drug used to treat chronic asthma and seasonal allergic rhinitis which may exist as four different stereoisomers, being the active one the (R)-*trans* isomer. A method separating Montelukast enantiomers and stereoisomers as well as its main degradation product was described [13].

This sensitive method allowed for a determination of a 0.02% of the enantiomeric and diastereoisomeric impurities and for a 0.01% of Montelukast main degradation product (Montelukast sulfoxide). Analysis of bulk drug and different dosage pharmaceutical formulations (pouches and chewable tablets) revealed content of Montelukast according to the label content [13].

Lansoprazole is a drug acting as inhibitor of the proton pump which is marketed as equally mixture of R- and S-lansoprazole [35]. Due to the increasing efforts in producing enantiopure drugs, and since R-lansoprazole has proven to stronger bind to human serum proteins than its S-enantiomer, chiral methods to ensure enantiomeric purity are desirable. On this aspect, Chung et al. developed a method based on a dual system consisting of sulfobutylether-β-CD and methyl-β-CD which enabled chiral separation for lansoprazole, where the enantiomer migration order was adequate, i.e., the enantiomeric impurity (S-enantiomer) migrated before the active enantiomer (R-lansoprazole). Content of S-lansoprazole was below the LOD (0.07%) in the six R-lansoprazole (dexlansoprazole) capsules analyzed [14].

Colchicine is an antiuremic drug commercialized as S-enantiomer due to its higher activity compared to its R-counterpart. Menéndez-López et al. screened different anionic CDs to find that succinyl-γ-CD and sulfated-γ-CD gave rise to best enantioresolution for colchicine [15]. Interestingly, a reversal in the EMO was found, the active principle was the first migrating enantiomer when using succinyl-γ-CD whereas that it migrated last with sulfated-γ-CD. This event was investigated using NMR and it is further detailed in the article [15]. Despite sulfated-γ-CD being the CD which enabled the enantiomeric impurity to migrate first, it did not enable its determination in 0.1% amounts due to its lower enantioresolution when compared to succinyl-γ-CD which enabled determination of levels as low as 0.1% (see Figure 2). Analysis of pharmaceutical formulations containing enantiomerically pure S-colchicine revealed that the enantiomeric impurity was below the LOD of the method in all cases [15]. Cinalcet is a chiral drug used for hyperparatiroidism in dialysis patients. Since the R-enantiomer is up to 75 times stronger than its S-enantiomer [36], chiral methodologies are needed. 2-HP-γ-CD as chiral selector enabled the chiral separation of cinalcet in tablets, in which amounts of the enantiomeric impurity were below the 0.1% [16].

One of the attractive features of CE is its ability to perform fast analysis, especially if analysis is carried out by the so-called short-end of the capillary this can be easily done if enantioresolution is good enough. This is the case of a recent work published by Szabo et al. based on chiral separation of praziquantel using negatively charged sulfated-β-CD [17]. Praziquantel is a drug used to treat the tropical disease called schistosomiasis. Despite being commercialized as racemate, only the R-enantiomer possesses the activity, being the S-enantiomer not only inactive but it also doubles the size of the tablet, has higher cytotoxicity and has bitter taste. Commercialization of enantiopure R-praziquantel is undergoing. The developed method was validated for determination of 0.1% of the enantiomeric purity under 10 min analysis and was further applied to analysis of two commercial tablets containing racemic praziquantel [17].

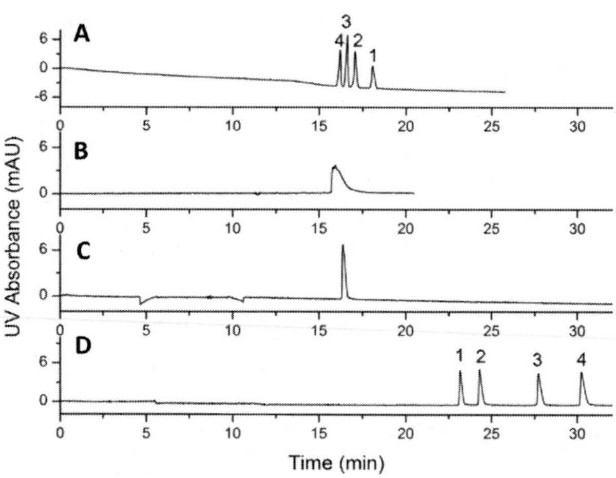

Figure 1. EKC separation of sitafloxacin stereoisomers standards using (A) 20 mg/mL highly sulfated γ-CD, (B) 20 mM γ-CD, (C) 10 mM Cu2+ and 10 mM D-Phe, and (D) 10 mM Cu2+, 10 mM D-Phe and 20 mM γ-CD. Peaks: (1) stereoisomer RSR (impurity), (2) stereoisomer SSR (impurity), (3) stereoisomer RRS (impurity), and (4) stereoisomer SRS (sitafloxacin). Reproduced with permission from [12].

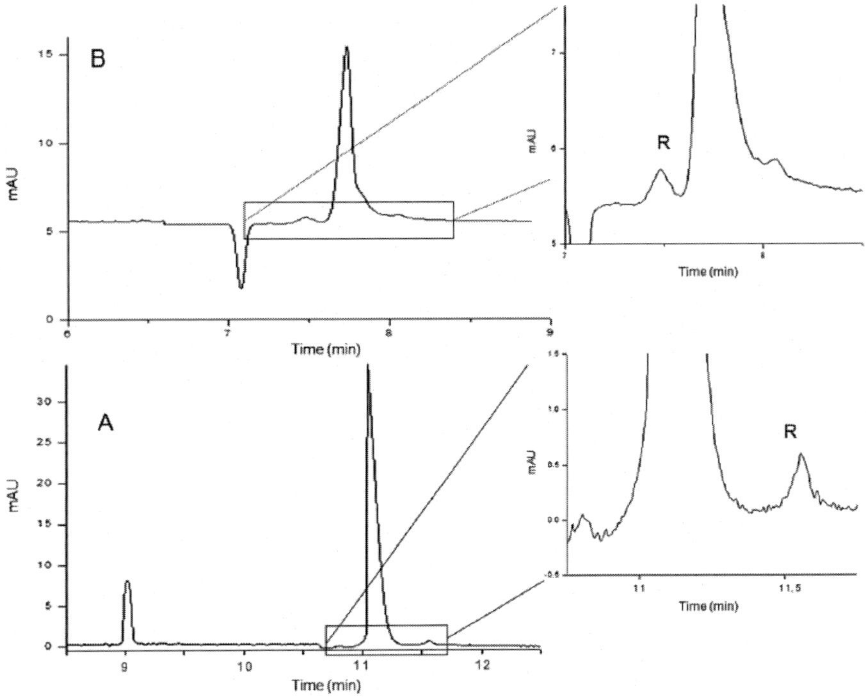

Figure 2. Electropherograms corresponding to the LOD of antiuremic drug colchicine for succinyl-γ-CD (A) or sulfated-γ-CD (B). BGE: 7 mM succinyl-γ-CD in 50 mM borate (pH 9.0) in (A) and 17 mM sulfated-γ-CD in 25 mM borate (pH 9.0) in (B). Reproduced with permission from [15].

Although the vast majority of chiral CE reports included in this chapter are based on aqueous CE, also NACE works have been described [18, 19]. In the first article, authors used a diacetone-D-mannitol-boric acid complex in the methanol-based BGE to enantiomerically resolve several β-agonists [18]. Subsequently, clenbuterol enantiomers were determined in three different racemic formulations (oral solutions) in which the measured content was in agreement with label content [18]. In the second contribution, authors developed a chiral method to separate 14 amino alcohol drugs employing lactobionic acid-boric acid and D-(+)-xylose-boric acid complex in a BGE containing methanol and triethylamine. Amounts quantified in propranolol tablets were in agreement with labelled content [19].

Finally, Li et al. demonstrated the feasibility of liposome electrokinetic chromatography (LEKC), a variety of EKC, in the chiral analysis of drugs naproxen, warfarin, ketoprofen and amlodipine [20]. The LEKC applied was based on using phosphatidylcholine and cholesterol as pseudo-stationary phase whereas that sulfobutyl ether-β-CD acted as chiral selector. Liposomes may provide more types of intermolecular interactions when compared to other pseudo-stationary phases such as micelles or micro emulsions [37]. Application was conducted towards analysis of a S-naproxen bulk sample, in which 0.9% of R-enantiomeric impurity was found [20].

Determination of Drugs in Biological Samples

CE has proven track-record to perform analysis of biological samples due to its inherent minimal need for sample (only few nLs are injected in a typical CE system), additionally to its other advantages already described. As can be seen in Table 2, biological samples described in this Chapter encompasses plasma, urine, and hair, among others. Unlike pharmaceutical samples, in biological samples more tedious sample treatments are required such as SPE or LLE, among others, mainly focused on getting rid of interfering molecules, especially when non-specific detectors are used such as UV-Vis detectors. MS detectors would be more suitable detection systems in biological sample analysis but, as earlier commented, MS has not been used in the articles published in this period. Most papers were focused on the quantification of drugs content in real samples although others performed the analysis of spiked samples. As in the pharmaceutical samples, a wide variety of drugs have been reported in the analysis of biological samples, and are discussed below.

Chlorpheniramine is an antihistamine drug whose S-enantiomer possesses most of the activity whereas that the R-enantiomer has sedative effects [38]. In a recent paper, the authors developed a strategy based on EKC-UV using cation-selective exhaustive injection (CSEI) sweeping, which enabled to improve the sensitivity up to 190 times compared to normal EKC obtaining LOQs of 0.025 μg/mL for chlorpheniramine enantiomers [21]. This sensitive approach was tested in a pharmacological study based on administrating racemic chlorpheniramine to rats. In the analysed plasma, S-enantiomer concentration was slightly higher than its R-counterpart, which, as the authors state, is in agreement to previous reports [39, 40].

Tranylcypromine is a monoamine oxidase inhibitor used as antidepressant drug. Using a surfactant-assisted electromembrane extraction technique combined with an enantioseparation method based on CDs, the authors extracted and enantioseparated tranylcypromine in spiked urine samples with recoveries around 85% [22]. The use of the surfactant-assisted electromembrane extraction technique is encouraged given the fact that it is a non-expensive and fast tool useful in the extraction and preconcentration of a wide array of metabolites [41].

Tetrabenazine is a drug used to treat hyperkinetic movement disorders as well as aiding in the therapy of Huntington's disease [42]. In a contribution, the authors developed a strategy based on dispersive liquid-liquid microextraction (DLLME) and CE to study the *in vitro* enantioselective metabolism of tetrabenazine in human liver microsomes [23]. As the authors point out, these studies support the pharmacokinetic studies found in literature.

A research including the development of a method to separate a new basic drug possessing two chiral centers has been reported [24]. This pyroglutamic acid derivative was undergoing drug development in order to be commercialized as a single enantiomeric drug as SR-isomer due to its pharmacological activity as immunomodulator for cancer treatment. Stress study based on submitting the drug to basic hydrolysis at 80°C during 24 h led to a significant degradation (30%) of the SR isomer mainly to the SS isomer. Storage of samples at drastic conditions (6 months at 40°C and 75% humidity) showed only a slightly conversion of the active enantiomer (SR) to the SS diastereoisomer, this means inversion occurred only at one of the chiral centers. CSEI sweeping improved the sensitivity 250 times when compared to sweeping (see Figure 3A). Finally, the method was used to assess possible *in vivo* interconversion in rats. Analysis of rat plasma samples, showed that presence of undesired impurities was negligible as a result of metabolism (Figure 3B) [24].

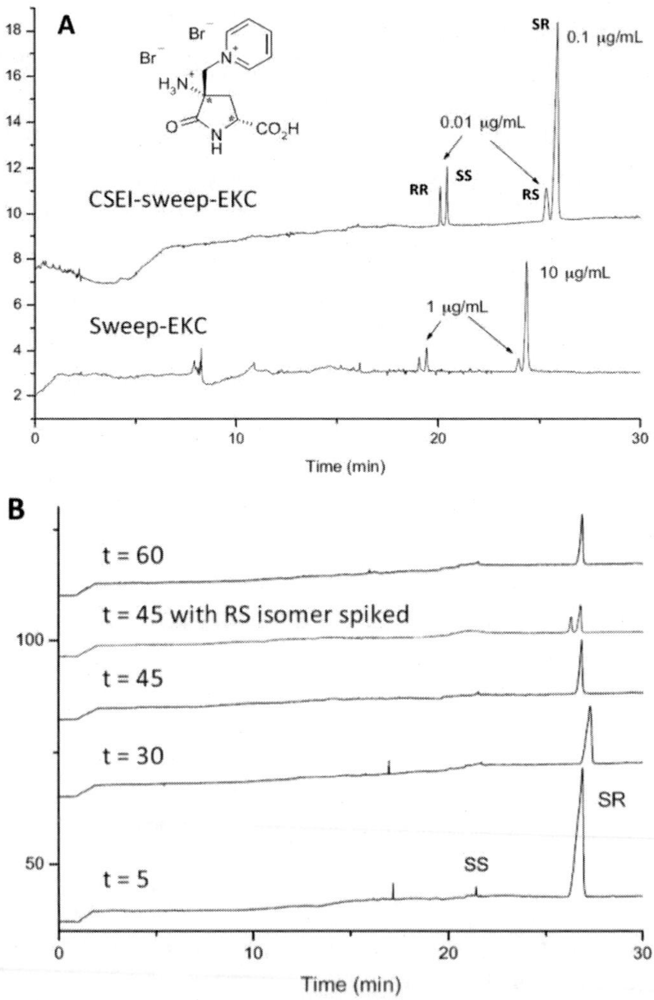

Figure 3. Comparison of CSEI-sweeping and only sweeping for an anti-cancer pyroglutamic acid derivative under development (A) and the *in vivo* pharmacokinetic study in rat plasma samples taken at 5, 30, 45 and 60 min upon drug administration (B). BGE: 2.5% (w/v) highly sulfated-β-CD in 50 mM phosphate buffer (pH 2.5). Reproduced with permission from [24].

As in the previous research [24], the study of the enantiomers of under-development drugs is necessary. In this aspect, researchers developed a method to enantioresolve two newly patented aminoalkanol derivatives which potentially present anticancer activity [25]. Recovery studies in spiked serum samples revealed values near 100% in all cases so this method demonstrates to be fit for purpose.

Ketamine is an anaesthetic drug which is used as S-enantiomer given its higher activity. In last two years, ketamine enantioselective metabolism has attracted much attention, being object of study in a total of four contributions, all of them using highly sulfated γ-CD in concentrations ranging from 0.1 to 0.66% in acid pH as the common chiral media [26-29]. Particularly, Sandbaumhuter et al. investigated the effect of four α2-receptor agonists (medetomidine, detomidine, xylazine, and romifidine) on the *in vitro* ketamine metabolization to norketamine and in the formation of metabolites 6-hydroxynorketamine and 5,6-dehydronorketamine in equine liver microsomes [26]. Results demonstrated that these four α2-receptor agonists severely impact ketamine metabolism, being medetomidine the one inhibiting the most. In addition, the authors suggest stereoselectivity occurring in the hydroxylation of norketamine, however it was not statistically significant, and researchers indicate they are currently working on it to obtain a better knowledge on this stereoselectivity event [26]. Another work devoted to the enantioseparation of ketamine and three of its metabolites (norketamine, 5,6-dehydronorketamine and 6-hydroxynorketamine), was also reported [27]. Therein, researchers demonstrated the enantioselective metabolism of ketamine in dog plasma. Since ketamine is often combined with other sedatives such as medetomidine as a mean to compensate for ketamine side effects, Sandbaumhüter et al. also investigated the metabolism of ketamine enantiomers and their metabolites in dogs in presence of sedatives [28]. Plasma from dogs, either sedated with medetomidine or anaesthetized with sevoflurane which were further treated with racemic ketamine or S-ketamine, was analyzed. Among the interesting findings detailed in the work, it is worth-highlighting that, upon treatment with medetomidine, both norketamine enantiomers were at a lower content in these plasma samples than in those from dogs treated with sevoflurane. Differences in stereoselectivity was found in the case of 6-hydroxynorketamine and 5,6-dehydronorketamine, whereas that it was not found for ketamine and norketamine [28]. Another strategy based on the determination of ketamine and norketamine in human hair samples was developed by Porpiglia et al. [29]. Therein, the authors used LLE based on hexane/ethyl acetate to extract ketamine and norketamine from human hair samples. In order to improve sensitivity the authors applied a field amplified sample stacking (FASS) strategy, enhancing the sensitivity up to 1000 times when compared to normal hydrodynamic injection, finally reaching LODs of 0.08 ng/mg for each enantiomer. In this work, recoveries in spiked human hair samples ranged from 49 to 91% for the four enantiomers. Hair samples of subjects who had consumed ketamine in the past showed the presence of ketamine and norketamine enantiomers [29].

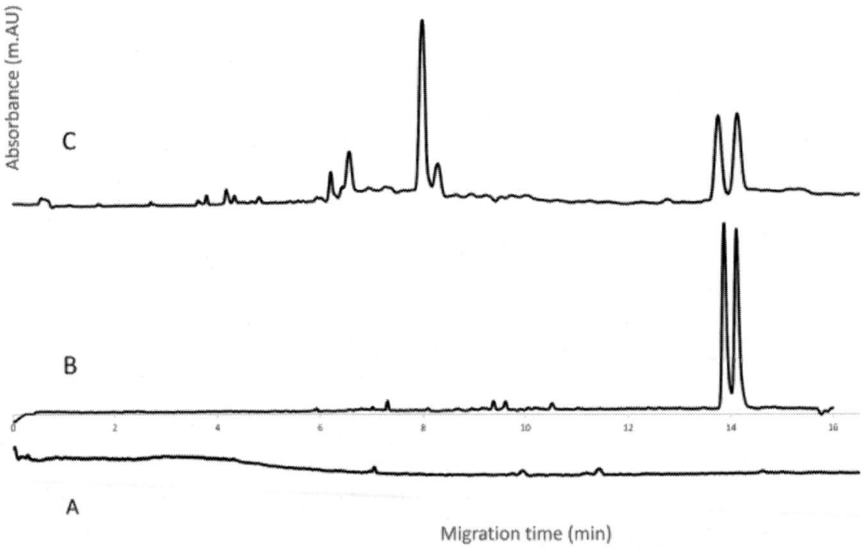

Figure 4. EKC electropherogram of enantioseparation of methadone in exhaled breath condensate (EBC). (A) Blank EBC, (B) EBC spiked with 2.5 μg/mL methadone enantiomers and (C) EBC sample of a patient under methadone maintenance therapy. BGE: 0.8% carboxymethyl-β-CD in 150 mM phosphoric acid-triethylamine (pH 2.5) containing 30% methanol. Reproduced with permission from [30].

Methadone is a prescribed drug to treat drug addiction and also used as pain reliever. Despite the 50-fold stronger opioid activity of the R-enantiomer over the S-enantiomer, this drug is administered as racemate. In last years, methadone has been studied in two contributions [30, 31]. In the first article, researchers developed a chiral CE method to determine methadone enantiomers in exhaled breath condensate (EBC). EBC is a non-invasive sampling which has demonstrated to be an interesting alternative to other invasive methods such as biofluids [30]. The authors validated the method according to the US FDA guidelines and EBC samples of patients under methadone maintenance therapy were analyzed. No significant differences were found in concentration of R- and S-methadone in EBC samples meaning that no stereoselective penetration from the blood flow to the lung lining fluid occurred (Figure 4) [30].

However, methadone concentration in EBC did poorly correlate with the ones found in urine and serum, possibly due to humidity and temperature of EBC sampling collection area, pH, individual habits of subjects, and drug consumption, among other factors, as researchers suggest [30]. In a second contribution, Baciu et al. developed an in-line SPE-EKC method including a previous sample treatment employing pressurized liquid extraction (PLE) in order to determine drugs of abuse, including methadone enantiomers, in hair [31]. Hair is, together with EBC another non-invasive sampling method. Additionally, since human hair grows typically at 1 cm/month, it provides a source of information of an individual's activity such as drug consumption. In fact, when hair samples from four subjects undergoing methadone maintenance treatment programme were analyzed, both S- and R-methadone were observed [31].

Cathinones, such as mephedrone and its metabolites, 4-methylephedrine and methylenedioxypyrovalerone (MDPV), are recreational drugs displaying structural similarities to amphetamines. Since these drugs are chiral, their enantiomers can display potent different pharmacological activity, thus, the presence of the enantiomers of illegal drugs is of utmost importance as to interpreting drug testing results. Baciu et al. developed a chiral method to enantioseparate mephedrone, 4-methylephedrine, and MDPV in spiked hair samples using an in-line SPE-CE setup [32].

Recoveries above 80% were obtained in all cases for the studied analytes including their enantiomers.

CONCLUSION

This chapter summarizes the most recent works concerning the chiral analysis of drugs in real samples (pharmaceutical formulations and biological samples) using CE. Without a doubt CE meets all the requirements to correctly carry out chiral analysis of drugs, obtaining low RLODs, following the ICH guidelines (detection of 0.1% of the distomer in most cases). Despite the added complexity in the analysis of drugs having more than a chiral center, CE has offered good resolution for both enantiomers and diastereoisomers, thanks to the advantages derived from its high efficiency.

An increasing interest was observed in using experimental design to better study the influence of different variables on a given chiral separation being the QbD principles employed in several of the works reported in the last years.

The most used strategy was EKC employing CDs, most particularly using negatively charged sulfated CDs. Also, dual selector systems based on a neutral and a negatively charged selector have been popular in last years. In addition to the chiral selector or mixtures of chiral selectors, micelles and liposomes were also added in some cases to the BGE to improve the enantiomeric resolution of drugs investigated using the so-called MEKC and LEKC working modes, respectively, being non-aqueous media (NACE mode) also employed although in very few occasions.

Even though the sensitivity is limited, detection based on UV-Vis demonstrates to perform well for the analysis of chiral drugs in pharmaceutical formulations as well as in biological samples. Sensitivity required was obtained using off-line (SPE, PLE, surfactant assisted electromembrane extraction technique or DLLME), in-line (SPE), and in-capillary (CESI-sweeping, FASS) preconcentration techniques. Some of these strategies originated an enhancement of the sensitivity up to 1000 times.

Future trends should be redirected towards reducing analysis time to make chiral analysis more straight-forward, as well as using chiral selectors which enable the distomer to be the first migrating enantiomer and to make more cost-effective some sensitive and selective detectors such as mass spectrometers to be more available to drug analysis laboratories.

ACKNOWLEDGMENTS

The authors thank financial support from the Spanish Ministry of Economy and Competitiveness (MINECO) (project CTQ2016-76368-P) and E. S. L. thanks the Comunidad of Madrid (Spain) and European funding from FEDER program (project S2013/ABI-3028, AVANSECAL-CM) for her contract.

REFERENCES

[1] Chankvetadze, B. (2002). Enantiomer migration order in chiral capillary electrophoresis. *Electrophoresis* 23:4022-4035.

[2] Stavrou, I. J., Agathokleous, E. A. & Kapnissi-Christodoulou, C. P. (2017). Chiral selectors in CE: Recent development and applications (mid-2014 to mid-2016). *Electrophoresis* 38:786-819.

[3] Zhu, Q. & Scriba, G. K. E. (2016). Advances in the Use of Cyclodextrins as Chiral Selectors in Capillary Electrokinetic Chromatography: Fundamentals and Applications. *Chromatographia*, 79:1403-1435.

[4] Saz, J. M. & Marina, M. L. (2016). Recent advances on the use of cyclodextrins in the chiral analysis of drugs by capillary electrophoresis. *Journal of Chromatography A*, *1467*, 79-94.

[5] El Deeb, S., Wätzig, H., El-Hady, D. A., Sänger-van de Griend, C. & Scriba, G. K. E. (2016). Recent advances in capillary electrophoretic migration techniques for pharmaceutical analysis (2013–2015). *Electrophoresis* 37:1591-1608.

[6] Zhu, Q. & Scriba, G. K. E. (2018). Analysis of small molecule drugs, excipients and counter ions in pharmaceuticals by capillary electromigration methods – recent developments *Journal of Pharmaceutical and Biomedical Analysis* 147:425-438.

[7] Niedermeier, S. & Scriba, G. K. E. (2017). A quality by design-based approach to a capillary electrokinetic assay for the determination of dextromepromazine and levomepromazine sulfoxide as impurities of levomepromazine. *Journal of Pharmaceutical and Biomedical Analysis* 146:402-409.

[8] Krait, S., Heuermann, M. & Scriba, G. K. E. (2018). Development of a capillary electrophoresis method for the determination of the chiral purity of dextromethorphan by a dual selector system using quality by design methodology. *Journal of Separation Science* in press (doi: 10.1002/jssc.201701166).

[9] Orlandini, S., Pasquini, B., Caprini, C., del Bubba, M., Dousa, M., Pinzauti, S. & Furlanetto, S. (2016). Enantioseparation and impurity determination of ambrisentan using cyclodextrin-modified micellar electrokinetic chromatography: Visualizing the design space within quality by design framework. *Journal of Chromatography A* 1467: 363-371.

[10] Krait, S., Dousa, M. & Scriba, G. K. E. (2016). Quality by design-guided development of a capillary electrophoresis method for the chiral purity determination of ambrisentan. *Chromatographia* 79: 1343-1350.

[11] Kazsoki, A., Fejos, I., Sohajda, T., Zhou, W., Hu, W., Szente, L. & Beni, S. (2016). Development and validation of a cyclodextrin-modified capillary electrophoresis method for the enantiomeric separation of vildagliptin enantiomers. *Electrophoresis* 37:1318-1325.

[12] Meng, R. & Kang, J. (2017). Determination of the stereoisomeric impurities of sitafloxacin by capillary electrophoresis with a dual chiral additives. *Journal of Chromatography A* 1506:120-127.

[13] Flor, S., Juan, M. H., Tripodi, V. & Lucangioli, S. (2016). Development of an enantioselective capillary electrophoretic method for the simultaneous determination of montelukast enantiomeric and diastereoisomeric forms and its main degradation product. *Electrophoresis* 37:2420-2428.

[14] Chung, H. K., Truong, Q. K., Mai, X. L., Choi, Y., Kang, J. S., Mar, W. & Kim, K. H. (2017). Determination of S-(2)-lansoprazole in dexlansoprazole preparation by capillary zone electrophoresis. *Archives of Pharmacal Research* 40:962-971.

[15] Menéndez-López, N., Valimaña-Traverso, J., Castro-Puyana, M., Salgado, A., García, M. A. & Marina, M. L. (2017). Enantiomeric separation of the antiuremic drug colchicine by electrokinetic chromatography. Method development and quantitative analysis. *Journal of Pharmaceutical and Biomedical Analysis* 138:189-196.

[16] Ginterová, P., Znaleziona, J., Knob, R., Dousa, M., Petr, J. & Sevcik, J. (2016). Enantiomeric purity control of -cinacalcet in pharmaceutical product by capillary electrophoresis. *Chemical Papers* 70:1024-1030.

[17] Szabo, Z. I., Gal, R., Szocs, L., Ludmerczki, R., Muntean, D. L., Noszal, B. & Toth, G. (2017). Validated capillary electrophoretic method for the enantiomeric quality control of R-praziquantel. *Electrophoresis* 38:1886-1894.

[18] Lv, L., Wang, L., Li, J., Jiao, Y., Gao, S., Wang, J. & Yan, H. Enantiomeric separation of seven β-agonists by NACE—Study of chiral selectivity with diacetone-D-mannitol–boric acid complex. *Journal of Pharmaceutical and Biomedical Analysis* 145:399-405.

[19] An, N., Wang, L. J., Zhao, J. J., Lv, L. L., Wang, N. & Guo, H. Z. (2016). Enantioseparation of fourteen amino alcohols by nonaqueous capillary electrophoresis using lactobionic acid/D-(+)-xylose-boric acid complexes as chiral selectors. *Analytical Methods* 8:1127-1134.

[20] Li, X., Du, Y., Feng, Z., Sun, X. & Huang, Z. (2017). A novel enantioseparation approach based on liposome electrokinetic capillary

chromatography. *Journal of Pharmaceutical and Biomedical Analysis* 145:186-194.

[21] Yao, Y., Zhou, L., Li, M. & Guo, X. (2018). The cation-selective exhaustive injection and sweeping capillary electrophoresis method for the analysis of chlorpheniramine enantiomers in rat plasma. *Journal of Pharmaceutical and Biomedical Analysis* 148:142-148.

[22] Fakhari, A. R., Kosalar, H. M., Asadi, S. & Hasheminasab, K. S (2017). Surfactant-assisted electromembrane extraction combined with cyclodextrin-modified capillary electrophoresis for the separation and quantification of tranylcypromine enantiomers in biological samples." *Journal of Separation Sciences* 41:475-482.

[23] Bocato, M. Z., de Lima Moreira, F., Perez de Albuquerque, N. C., Masetto de Gaitani, C. & Moraes de Oliveira, A. R. (2016). *In vitro* enantioselective human liver microsomal metabolism and prediction of *in vivo* pharmacokinetic parameters of tetrabenazine by DLLME-CE. *Journal of Pharmaceutical and Biomedical Analysis* 128:528-537.

[24] Martínez-Girón, A. B., Marina, M. L. & Crego, A. L. (2017). Chiral separation of a basic drug with two chiral centers by electrokinetic chromatography for its pharmaceutical development. *Journal of Chromatography* 1467:427-435.

[25] Grodner, B., Lukaszkiewicz, J. & Napiorkowska, M. (2016). Determination of the stereoisomers in aqueous medium and serum and validation studies of racemic aminoalkanol derivatives of 1,7-dimethyl-8,9-diphenyl-4-azatricyclo[5.2.1.02,6] dec-8-ene-3,5,10-trione, potential new anticancer drugs, by capillary electrophoresis. *Journal of Separation Sciences* 39:3246-3253.

[26] Sandbaumhüter, F. A., Theurillat, R., Bettschart-Wolfensberger, R. & Thormann, W. (2017). Effect of the α2-receptor agonists medetomidine, detomidine, xylazine, and romifidine on the ketamine metabolism in equines assessed with enantioselective capillary electrophoresis. *Electrophoresis* 38:1895-1904.

[27] Theurillat, R., Sandbaumhüter, F. A., Bettschart-Wolfensberger, R. & Thormann, W. (2016). Microassay for ketamine and metabolites in plasma and serum based on enantioselective capillary electrophoresis

with highly sulfated γ-cyclodextrin and electrokinetic analyte injection. *Electrophoresis* 37:1129-1138.

[28] Sandbaumhüter, F. A., Theurillat, R., Bektas, R. N., Kutter, A. P. N., Bettschart-Wolfensberger, R. & Thormann, W. (2016). Pharmacokinetics of ketamine and three metabolites in Beagle dogs under sevoflurane vs. medetomidine comedication assessed by enantioselective capillary electrophoresis. *Journal of Chromatography A* 1467:436-444.

[29] Porpiglia, N., Musile, G., Bortolotti, F., de Palo, E. F. & Tagliaro, F. (2016). Chiral separation and determination of ketamine and norketamine in hair by capillary electrophoresis. *Forensic Science International* 266:304-310.

[30] Hamidi, S., Khoubnasabjafari, M., Ansarin, K., Jouyban-Gharamaleki, V. & Jouyban, A. (2017). Chiral separation of methadone in exhaled breath condensate using capillary electrophoresis. *Analytical Methods*, 9:2342-2350.

[31] Baciu, T., Borrull, F., Aguilar, C. & Calull, M. (2016). Findings in the hair of drug abusers using pressurized liquid extraction and solid-phase extraction coupled in-line with capillary electrophoresis. *Journal of Pharmaceutical and Biomedical Analysis* 131:420-428.

[32] Baciu, T., Borrull, F., Calull, M. & Aguilar, C. (2016). Enantioselective determination of cathinone derivatives in human hair by capillary electrophoresis combined in-line with solid-phase extraction. *Electrophoresis* 37:2352-2362.

[33] Q3B, Impurities in new drug products, in: *International Conference on Harmonization. ICH Harmonized Tripartite Guidelines*, ICH, 2006.

[34] Robert, T. A., Hagardorn, A. N. & Daigneault, E. A. (1982). Differential stereoselectivity of methotrimeprazine enantiomers for selected central nervous system receptor types. *Molecular Pharmacology* 21:315–319.

[35] Nevado, J. J., Peñalvo, G. C., Sánchez, J. C., Mochón, M. C., Dorado, R. M. & Navarro, M. V. (2009). Optimisation and validation of a new

CE method for the determination of lansoprazole enantiomers in pharmaceuticals. *Electrophoresis* 30:2940-2946.

[36] Choulwar, A. K., Mungantiwar, A. A. & Chintamaneni, M. (2011). A comparative, bioequivalence study to evaluate the safety and pharmacokinetic profile of single dose cinacalcet hydrochloride tablets in healthy, adult, human subjects under fed conditions. *Annals of Biological Research* 2:35-50.

[37] Martinek, K., Levashov, A. V., Klyachko, N., Khmelnitski, Y. L. & Berezin, I. V. (1986). Micellar enzymology. *European Journal of Biochemistry* 155:453-468.

[38] Chen, Q. C., Jeong, S. J., Hwang, G. S., Kim, K. H. & Kang, J. S. (2008). Enantioselective determination of chlorpheniramine in various formulations by HPLC using carboxymethyl-β-cyclodextrin as a chiral additive. *Archives Pharmacal Research* 31:523–529.

[39] Hiep, B. T., Fernández, C., Tod, M., Banide, H., Thuillier, A., Lacour, B., Farinotti, R. & Gimenez, F. (2001). Intestinal absorption and metabolism of chlorpheniramine enantiomers in rat. *Chirality*, 13:207–213.

[40] Fried, K. M., Young, A. E., Yasuda, S. U. & Wainer, I. W. (2002). The enantioselective determination of chlorpheniramine and its major metabolites in human plasma using chiral chromatography on a β-cyclodextrin chiral stationary phase and mass spectrometric detection. *Journal of Pharmaceutical and Biomedical Analysis* 27:479–488.

[41] Lee, J., Lee, H. K., Rasmussen, K. E. & Pedersen-Bjergaard, S. (2008). Environmental and bioanalytical applications of hollow fiber membrane liquid-phase microextraction: a review. *Analytical Chimica Acta* 624:253-268.

[42] Chen, J. J., Ondo, W. G., Dashtipour, K. & Swope, D. M. (2012). Tetrabenazine for the treatment of hyperkinetic movement disorders: a review of the literature. *Clinical Therapeutics* 34:1487-1504.

In: Capillary Electrophoresis ... ISBN: 978-1-53615-223-4
Editors: O. Núñez and X. Subirats

Chapter 5

NEW TRENDS IN CAPILLARY ELECTROCHROMATOGRAPHY TECHNIQUES

Chiara Fanali[1,*] and Giovanni D'Orazio[2]
[1]Unit of Food Science and Nutrition, Department of Medicine, Università Campus Bio-Medico di Roma, Rome, Italy
[2]Institute of Chemical Methodologies, Italian Research Council (CNR), Monterotondo, Italy

ABSTRACT

Capillary electrochromatography (CEC) is an analytical technique performed in capillaries with a thin internal diameter containing a stationary phase (SP). Compounds are moved into the column to the detector by a strong electroosmotic flow (EOF) generated by the application of a high voltage (10-20 kV). The separation mechanism involved in the analysis process is based on both chromatographic and electrophoretic principle. Therefore it offers high efficiency and high selectivity typical of capillary electrophoresis and high performance liquid chromatography (HPLC), respectively. CEC has been widely studied and the literature covers several aspects related to this miniaturized technique

* Corresponding Author's Email: c.fanali@unicampus.it.

such as theory, instrumentation, selection and use of different stationary phases, applications etc.

The aim of this chapter is to present the main features of CEC and its potentiality for the separation of compounds of different nature (uncharged and charged species). Instrumentation currently used considering the detectors and the coupling with mass spectrometry (MS), optimization of experimental conditions, the different SPs have also been discussed. Finally some selected applications in the field of drug analysis, protein, enantiomers, nutraceutical compounds separation are described.

Keywords: capillary electrochromatography, CEC, mass spectrometry, CEC-MS, stationary phases, monolithic, electroosmotic flow

INTRODUCTION

Capillary electrochromatography (CEC) is a liquid chromatographic (LC) technique performed under the application of a relatively high electric field. It combines the advantages of both capillary electrophoresis (CE) and high performance liquid chromatography (HPLC), namely high efficiency and selectivity, respectively. The driving force involved in the CEC process, responsible for the transport of analytes along the column to the detector, is produced by a strong electroosmotic flow (EOF). EOF is generated by the application of a high voltage in presence of charged groups on the capillary wall and on the stationary phase (SP) surface. It has been reported that the EOF has a flat flow profile, that unlike that observed in HPLC (parabolic), is considered the main reason for the higher chromatographic efficiency achieved in CEC. The extent of the EOF is influenced by several experimental parameters such as type and charge of the particles, applied voltage, length of the column, mobile phase (MP) composition (buffer, ionic strength, pH, organic solvent) etc.

Concerning the SP used in CEC, in principle, any type of SP employed in HPLC could be used with this electromigration technique. However it must be kept in mind that charged or ionizable groups must be present. The SP is usually contained into capillaries of thin diameter (10-100 μm I.D.) either coated or bonded on the wall, on silica or other material particles or on polymeric material (monoliths). The capillary, containing the SP, is positioned into a capillary electrophoretic instrumentation and after sample injection, a high voltage applied (2-30 kV) for the CEC run.

Separated compounds are usually detected utilizing UV or fluorescence detectors. However for mass determination and/or characterization, mass spectrometry (MS) has also been used.

In this chapter an overview of CEC, considering general principles, instrumentation, hyphenation with MS, method optimization and recent applications reported in literature will be illustrated.

Principles of Capillary Electrochromatography

CEC is an analytical technique where liquid chromatography is combined with electrophoresis. The advantages of the two techniques are beneficially used to obtain a high efficiency and excellent selectivity, typical of CE and HPLC, respectively. In CEC both analytes and mobile phase are transported to the detector by an EOF that is originated by the application of a relatively high electric field to the capillary containing a selected electrolyte in contact with charged groups present on the capillary wall and/or on the stationary phase surface. On the surfaces, a diffuse double layer of electrolyte ions with opposite charge to that of the surface is formed and is influenced by several parameters. Among them, applied electric field, type and charge of the SP, electrolyte concentration and ionic strength, organic additive type and concentration are the most important for a correct selection of the experimental conditions. Interesting information about the EOF have been reported by Pretorius et al. in 1974 [1] comparing the results achieved utilizing both CEC and LC. Later on, detailed theoretical studies were carried out by Knox's group [2, 3] and Rathore and Horvath [4].

Some review papers related to the applications done with CEC are available in literature [5-8]. In addition, theoretical consideration have also been reported [9, 10].

The driving force involved in CEC, generate a mobile phase flow with a flat profile, in contrast with the one observed with LC miniaturized techniques (e.g., nano-LC) with a parabolic profile. Therefore this feature is on the base of the higher efficiency that can be obtained in CEC. In addition EOF is not influenced by the particles diameter and for that reason smaller particles can be advantageously used for obtaining higher efficiency (A and C terms in van Deemter equation are lower in CEC than in nano-LC). Finally it has been reported that the use of high porous material enhances the EOF [11]

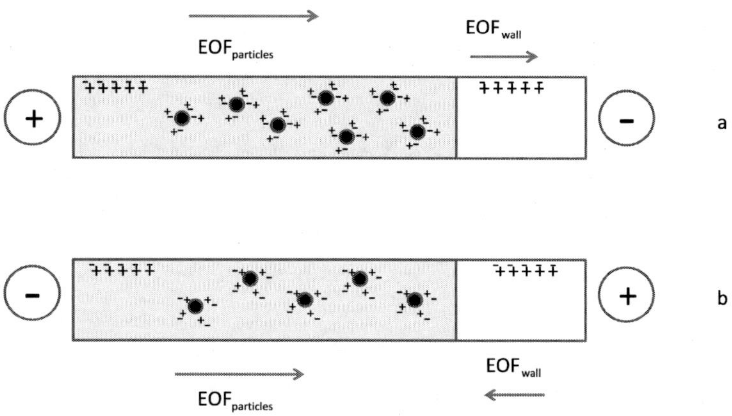

Figure 1. Scheme of two capillaries packed with a) negatively and b) positively charged particles generating opposite electroosmotic flow.

Figure 1 shows a scheme of two packed capillaries employed in CEC. As can be observed in fig. 1a, capillary wall and particles are negatively charged and a double layer is formed with the positive ions of the mobile phase. Applying an high voltage, the positive ions, surrounded by solvent molecules, will move to the cathode generating a strong EOF. The resulting flow is given by the sum of the EOF generated by the particles and the capillary wall. Fig. 1b depict a different situation where the wall (silica based) is negatively charged, while particles are positive. Therefore the EOF generated by the particles, where on the outside double layer positively charges, will move to the anode. An opposite flow (of lower strengh) is also generated to the opposite side of the capillary. In this case (reversed EOF) can take place and can be used when charged silica particles (e.g., aminopropyl) are employed. The reported example can serve here to show that EOF in different directions and of different strength can be generated depending of the type of silica particles used. Similar observations can be done with other stationary phases employed, e.g., polymeric material.

Analytes separation is performed in a stationary phase (SP), present into a capillary (usually of fused silica) and can be either coated/bonded on the wall, on monolithic material or on silica particles [12].

Most of the separations performed by CEC have been carried out using an isocratic elution mode. However gradient mode has also been reported [13-16]. In order to produce a gradient flow, different approaches have been used, e.g., by simply changing the inlet vial containing a different mobile phase or mixing the mobile phases present in two capillaries connected to the column through a t union and under the voltage generated by two power supplies or generated by a micro-LC pump. Recently, an affinity gradient nano-stationary phase has also been reported for proteins separation in microchips [17].

Capillary Columns and Instrumentation Used in Capillary Electrochromatography

Capillary Columns

Capillary columns used in CEC are mainly based on fused silica material of thin internal diameter (I.D. in the range 10-100 μm). The column contains a selected stationary phase either coated/bonded on the wall or on particles surface or on the polymer (monolith). In general the same SPs employed in HPLC have been used in CEC, however charged or chargeable groups had to be present in order to generate the strong EOF. Therefore, based on the type of column used, CEC can be classified as: i) open-tubular (OT-CEC), ii) packed capillary (p-CEC) and ii) monolithic.

In OT-CEC the SP is positioned on the capillary wall either coated or bonded. Tsuda and co-workers firstly demonstrated the possibility to carry out OT-CEC in silica capillaries of 30 μm I.D. modified with octadecyl groups [18]. Some standard organic uncharged compounds have been separated under the influence of a EOF and interaction with the octadecyl groups present on the wall. Elution was done utilizing mixtures of acetonitrile-water (30:70, v/v). Later on very good enantiomers separation has been shown with thin silica capillary modified with various chiral selectors (CS). Mono-6-O-octamethylene-permethyl-β-cyclodextrin (Chirasil-Dex), thermally attached to the capillary wall was applied for some barbiturate enantiomers separation [19]. Permetylated cyclodextrins have also been used by Armstrong et al. [20] for hexobarbital enantiomers separation in 50 μm I.D. capillaries, while Francotte employed polysaccharides coated on the wall of a capillary of the same I.D. for the separation of gluthetimide enantiomers by OT-CEC [21]. In addition to the mentioned CS, it has been reported about the use of molecular imprinted polymers bonded to the capillary wall [22]. In this case the capillary I.D. was 100 μm and the SP thickness was 0.1 μm applied for the separation of 2-phenylpropionic acid enantiomers. Later-on some other SPs have been applied for OT-CEC, e.g., bonded polymethacrylate for the separation of p-hydroxybenzoates [23], proteins such as lysozyme for amino acids and mephenytoin enantiomers separation [24], bonded macrocyclic dioxopolyamine [25], adsorbed avidin [26], n-octadecyl- and cholesterol etched chemically bonded [27], calyx(6)crown ether capillary coated [28], bonded cyano pentoxy biphenyl liquid crystal with etching procedure [29], dodecanethiol gold nanoparticles (etched procedure) for the separation of some plyciclic aromatic hydrocarbon [30], phenylalanine functionalized tentacle-type polymer for proteins separation [31], thiolated β-CD modified gold nanoparticles (CD-GNPs) adsorbed on the wall for the chiral separation of selected drugs [32], negatively charged ghraphene adsorbed on the wall for the separation of hair dye samples [33], gold nanoparticles/thiols β-cyclodextrin coating for enantiomers resolution [34], polymer-modified fibrous mesoporous silica nanoparticles for polycyclic aromatic hydrocarbons (PAHs) and proteins [35].

Although good results have been shown, OT-CEC exhibited some limitations related to i) the inadequate interations between analytes and SP on the contrary of packed or monolithic capillaries and ii) small load capacity due to the small area for bonding the SP. The two problems have been studied along the year proposing, e.g., the increase of capillary surface utilizing an etching procedure [27] and employing longer and/or thinner capillaries (10 μm I.D.).

Most of the columns currently employed in CEC are the ones obtained by packing particles into capillaries, generally of fused silica. This type of capillary resulted to be the most useful mainly because the inert properties and because allow the detection directly. The most used SP includes silica-based material appropriately modified with various groups such as C18, C8, CN, phenyl, phenyl-hexyl, polysaccharides, glycopeptides antibiotics, cyclodextrins etc. However it is note worthy remember that charged or chargeable groups have to be present in order to generate the EOF. The silica modified with C18 particles (ODS phases) are the most used till now and commercially available. They are applied for the separation of a large number of compounds such as drugs, pesticides, peptides etc. In the field of enantiomers separation, silica-based polysaccharides, glycopeptides antibiotics and quinine are the most employed. These SPs make use of both porous and non-porous silica with particles of 3-5 μm diameter; however also sub-2 μm particles have been successfully used [36-38].

Although different packing procedures have been proposed [39, 40], the most popular makes use of the slurry packing method. This is based on the preparation of a slurry mixing the SP with an appropriate organic solvent, e.g., methanol, acetonitrile, acetone and then the mixture, under the effect of a ultrasonic bath, pumped into the capillary. At the end of the capillary, a mechanical frit is positioned. After packing the desired length of the capillary, the SP is removed and water pumped to eliminate the organic solvent. Inlet and outlet frits are prepared utilizing a heated wire at 600-700 °C for a few seconds (6-8 s). The column is equilibrated pumping the mobile phase and then, after preparing the detector window, positioned into a cartridge for the CEC experiments. Before CEC runs, the capillary is again equilibrated by applying different voltages until a stable signal is recorded. For CEC runs, external pressure, at the two electrode compartments, has to be applied to avoid bubble formation. Alternatively an external pressure only at the inlet capillary side could be applied. However in this case compounds separation is obtained not only by CEC but also by nano-LC with consequent decrease of efficiency. Different packing procedures have been proposed taking in mind the problems related to bubble formation. Tapered columns of about 10 μm, blocking the silica particles without frits have been successfully used [41]. In addition, single particle fritted capillary has been proposed obtaining robust packed columns for CEC separations [42].

Frits prepared with silica can also be replaced by monolithic material [43-45] without applying external pressure. This last approach resulted very advantageous when CEC was coupled with MS.

Alternatively to microparticulate packed column, monolithic beds without frits have been proposed. The material contains micropores and macropores with no interstitial volume very advantageous for faster diffusional mass transfer [46]. In addition a high flow rate, very advantageous when working with nano-LC, can be applied because the low backpressure generated [47]. Other advantages, when using monolithic beds, includes i) their in situ preparation that is very useful for example when chip columns have to be produced, ii) modeling the column for appropriate chromatographic interactions selecting the most useful chemical compounds.

Monolithic capillary column can be obtained by polymerization of either inorganic material, e.g., silica, zirconia, titania or different organic monomers (methacrylate, styrene-divynilbenzene) [48]. Before polymerization process, the capillary column is modified with specific chemical reaction in order to bond the beds on the wall. Afterwards the in situ monolithic columns preparation can be done with different methods including the use of selected monomers in presence of a porogen and initiator. The polymerization is obtained by i) free radical polymerization [49] or ii) photopolymerization [43]. Other types of monolithic columns consider the sol-gel technology based on the in situ preparation of an hydrogel that is converted to a xenogel after drying and then chemically modified with groups promoting chromatographic interaction [50].

Some review papers dealing with the synthesis and use of monolithic columns in CEC have been reported in the last five year, e.g., [8, 51-57]. During this considered time about 135 papers have been published focusing on the synthesis of new polymeric material and their use as columns for various CEC applications. Fumed silica nano-particles, modified with glucosamide to study hydrophilic interaction have been prepared [58]. Other examples of new monolithic modified material have been proposed, e.g., included ionic liquids or graphene oxide [59-64]; multi-walled or sigle-walled carbon nano-tube [65-68]; dopamine for imprinted monolithic columns [69]; cyclodextrins [57, 70]; gold nano-particles [71]; pepsin or bovine serum albumin [72-74]; lanthanides [75]; phosphorylcholine [76]; antibiotics [77]; epinephrine [78]. Finally zirconia-based monolithic columns have been successfully proposed, e.g., Dixit et al. synthesized a zirconia hybrid monolithic column also containing clindamycin by a sol-gel procedure for enantiomers separation of some acidic compounds by CEC. Carprofen, ketoprofen, flurbiprofen, ibuprofen, suprofen and warfarin enantiomers were base-line separated with a polar-organic mixture mobile phase [79]. The same group also used a similar hybrid monolithic column modified with quinine [80] or phosphate-β-cyclodextrin [81] for enantiomers separation of basic compounds by CEC.

Instrumentation, Conventional Detectors and Hyphenation with Mass Spectrometry

Instrumentation currently used for CEC experiments is usually the one employed for capillary electrophoresis (CE) analysis where relatively high voltages (10-30 kV) can be applied. Although various apparatuses are commercially available, not all of them can perform CEC run because an external nitrogen or pump pressure has to be applied. This is a need because, due to the presence of frits retaining particles, bubbles are generated during the CEC runs. This problem was less remarkable using monolithic columns where frits are not necessary. It has also been demonstrated that CEC without pressure assistance can be easily carried out with very good results preparing packed capillary with monolithic frits [45]. In other cases, employing silica packed particles a pressure can also be applied at the inlet only avoiding bubble formation. However this approach combines CEC and nano-LC introducing, in addition to the EOF, a parabolic flow typical of LC and consequently a lower efficiency has to be expected. The capillary is positioned into an appropriate cartridge after the preparation of a small window just close to the outlet frit or the end of the monolithic material for the UV/fluorescence detection. After equilibration of the stationary phase with the selected mobile phase, samples are injected. Injection is done either by electrokinetic (4-8 kV for 0.1-1 min) or applying a pressure (8-10 bar for 1-2 min). Then a voltage in the range 8-25 kV is applied for the CEC run. Although UV-Vis detector is the most used tool, other systems have been employed, e.g., laser induced fluorescence (LIF) [82], conductivity [83], amperometric [84], etc. Finally CEC has been coupled with mass spectrometry (MS). The use of this detector type is extremely important especially when real complex samples have to be analyzed. In fact it offers the possibility to acquire important information related to the separated analytes with possibility to determine their exact mass and identification. Most of the work published till now makes use of ion source mainly based on electrospray (ESI) MS with the need of dedicate interfaces. These interfaces are commercially available for both nano-LC and CE. The coupling of CE with MS is more difficult because an electric connection with the interface and the power supply is necessary. This is mandatory to perform the CE run and for analytes ionization. The most simple interface used in CE consists of the so called sheath liquid (SL) one [85-91].

Figure 2 shows a scheme of this coupling tool.

As can be observed the system contains a triaxial tube where the mobile phase and analytes are moving along the capillary, a selected sheath-liquid, moved by a pump (flow at 3-6 μL x min-1) connected with the power supply and a stream of nitrogen. The same interface has been used in our laboratory for the separation of non-steroidal anti-inflammatory drugs by CEC [91]. This interface presented some drawbacks for these purposes, e.g., reduced sensitivity, poor resolution, current instability. In addition relatively long packed capillaries had to be employed. The use of the SL, being at flow velocity higher than the mobile phase, caused turbulence at the end of the capillary producing a reduction of both chromatographic efficiency and resolution with also problems in the ion-spray stability. Therefore, later on CEC experiments were performed utilizing the liquid junction (LJ) interface [89] that was firstly applied to CE-MS [92, 93]. As can be seen in Figure 3, the capillary column join the tip into an electric vessel connected to a syringe containing a selected solvent (e.g., methanol and formic acid or buffer or ammonia). This solvent was delivered at low flow by gravity. In this way a stable spray was obtained, avoiding bubble formation that could interfere with current stability. In addition good efficiency and good resolution was obtained. The sensitivity was improved because no peak dilution took place and the whole mobile phase-analytes were introduced into the MS analyzer. Figure 4 shows as an example of the good results that can be achieved with this LJ interface using CEC for the separation of a basic enantiomers compounds mixture using a column packed with vancomycin bonded silica particles.

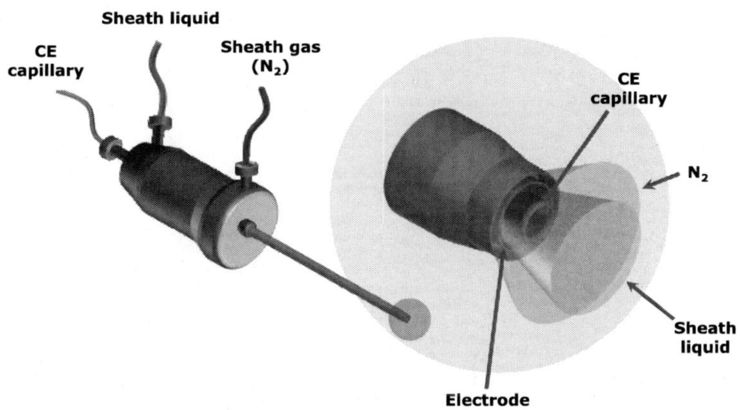

Figure 2. Scheme of co-axial sheath liquid interface for CE/CEC-MS system.

Later on a new SL interface was studied and applied for the separation by CEC-MS of some organophosphorous pesticides and some drugs. The same interface could be used for both nano-LC and CEC coupled with MS analysis. The interface was assembled utilizing a multiport fittings tee made of ULTEM® resin where the capillary column, the tip and the ionization liquid were connected for CEC experiments. A relatively low pressure (with nitrogen) was applied to the spray liquid. The results were quite satisfactory for both CEC and nano-LC obtaining very good efficiency, resolution, sensitivity and repeatability. In addition it was easy to use; no need for microscopy in order to optimize the distance between capillary column and tip as in the case of the previously prepared LJ interface [94] (see Figure 5).

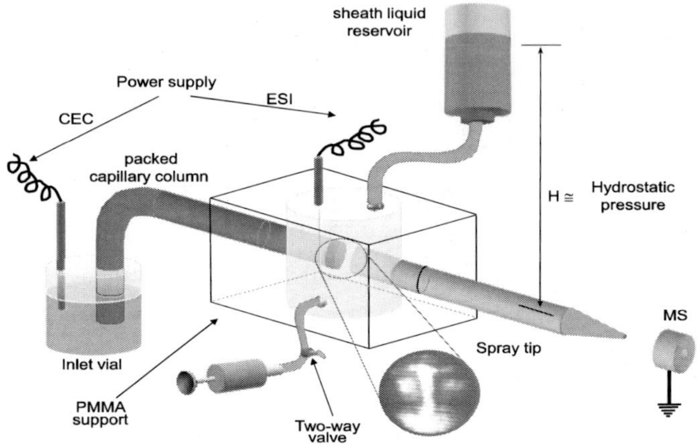

Figure 3. Scheme of pressurized liquid junction interface for CE/CEC-MS. Reprinted from [89] with permission of Elsevier.

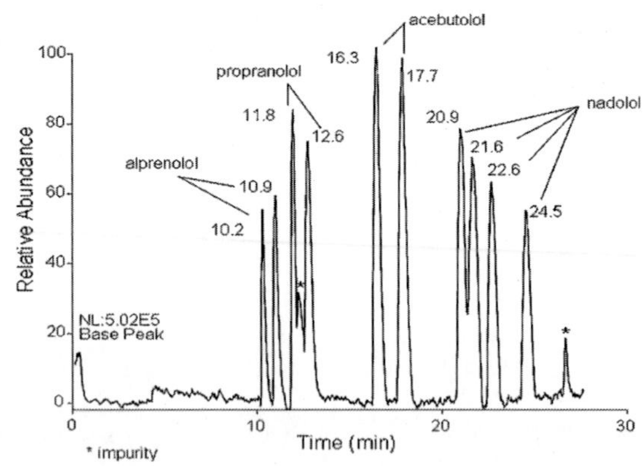

Figure 4. Enantiomeric separation of some standard drugs by CEC coupled with MS employing the LJ ESI interface. Capillary column: 100 µm I.D., 26.0 cm packed with (5µm) silica modified with vancomycin. Mobile phase: 100 mM NH4Ac pH ~ 6/MeOH-ACN, 1:9 (v/v); applied voltage, +20 kV. Electrokinetic injection: 12 kV for 5 s. The sheath-liquid was 0.5% (v/v) AcH in MeOH/H_2O 80:20 (v/v) ~2.5 kPa, applied hydrostatic pressure to outlet compartment only. Reprinted from ref. [89] with permission of Elsevier.

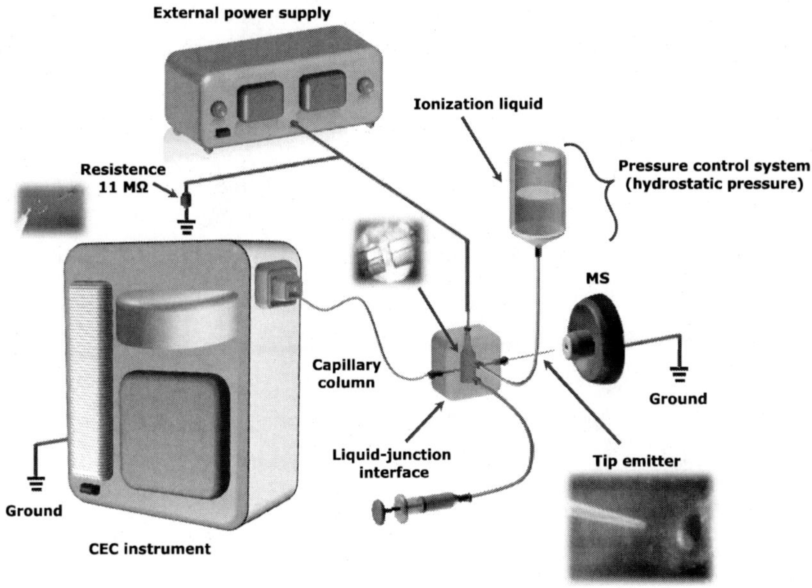

Figure 5. Scheme of a CEC-MS system using a LJ experimental interface.

Method Optimization in CEC

Method optimization is an important step in obtaining good results for CEC separation and is mainly based on the experience taking also in mind some important requirements. Among them, a strong EOF must be generated in order to have as short as possible the analysis time. As previously mentioned a stationary phase with charged/chargeable groups have to be selected together with appropriate experimental condition such as the mobile phase composition (buffer type and concentration, buffer pH, organic modified type and concentration, applied voltage etc.).

As an example, when performing CEC in the reversed elution mode using a silica C18 stationary phase with particles of 5 μm diameter, a capillary length of 20-25 cm is selected. In this case the charge of the SP stands in the residual silanol groups present on the silica and therefore the EOF is in the direction of the cathode. The selected mobile phase is usually composed by a mixture of acetonitrile/buffer (70:30 or 80:20, v/v). The buffer (ammonium or sodium acetate or phosphate) at pH 6-8 and concentrations 5-25 mM are selected. The high pH increases the EOF because the higher dissociation of silanols groups on the silica particles. The organic solvent must have a high dielectric constant and a low viscosity. Alternatively or in mixture to acetonitrile, other solvents have been used, e.g., methanol or ethanol or isopropanol. The use of these solvents, on one end modify the double layer formed on the silica surface changing the EOF, on the other hand can influence the electrophoretic mobility of charged analytes (if studied) and/or modify the selectivity. Modification of the double layer is also done changing the buffer concentration; lower concentrations increases the EOF. Finally CEC performed at higher voltage and higher temperature enormously reduce the analysis time by increasing EOF and electrophoretic mobility. However the use of high voltage also causes an increase of the generated current that could produce bubbles during CEC runs.

Some Selected Applications

CEC has been applied to the analysis of compounds belonging to various fields such as pharmaceutical, biochemical, food and agrochemical, forensic, environmental ones. Some review papers, dedicated to this topic, have been reported [7, 52, 54, 95-102].

A large number of studies related to CEC, available in literature, are dealing with the development of new capillary columns where standard compounds have been used as model mixture. Among the novel SP proposed, silica modified particles with C8, C18, bidentate C18, CN, mixed mode (SCX-C18), vancomycin, teicoplanin, amylose or cellulose modified polysaccharides etc. have been employed in packed capillaries.

CEC analysis of drugs is a good example of the application of the reversed phase mechanism using C18 or C8 silica columns. Ketorolac and some analogous have been quantified [103], while three analgesic compounds (caffeine, paracetamol and acetylsalycilic acid) were determined in tablets [104], respectively. The same chromatographic mechanism was also applied to the analysis of: vegetable oils (vitamins as tocopherols) [105] or extra virgin olive oils (phenolic compounds) [106] achieving good results with column packed with particles of 3 or 4.2 μm diameter, respectively; cogent C18 SP was also applied to the analysis of tea (catechins and methylxanthines) [107] and the analysis of 5-nitroimidazole antibiotic spiked in urine samples - here a column packed with Cogent bidentate C18 and silica particles (3:1, w/w) has been used for increasing the EOF and also coupling with MS [108]; pepper extracts (aromatic terpenic compounds) [109]. Vegetable oils, by reversed phase SP, were also analyzed by Abidi et al. [110] (sterols and related steryl esters) and Dermaux et al. [5] (free fatty acids). Other applications include also the analysis of vegetables [111, 112]; licorice (nucleosides and nucleobases) [113]; biogenic amines in foods after derivatization with 4-fluoro-7-nitro-2,1,3-benzoxadiazole using laser induced fluorescence detector [114] and infant food (carotenoids) where a capillary packed with C30 SP was used [115].

Some applications have been carried out by CEC utilizing chiral stationary phases where the separation mechanism is based on affinity interactions, e.g., modified polysaccharides or glycopeptides antibiotics silica based adsorbed/bonded or bonded, respectively. The interactions involved in the process include, e.g., hydrogen bonding, π-π, electrostatic, hydrophobic and in some cases inclusion complexation (vancomycin) etc. Cyclodextrins or their derivatives have also been used; here the inclusion complexation mechanism is mainly involved in the enantioseparation mechanism.

In the case of polysaccharides such as amylose or cellulose, they are usually modified with carbamate containing groups like methyl and chlorine. The position of these groups strongly influences the enantioresolution modifying the properties of the aromatic ring. The good enantioselectivity of amylose tris(5-chloro-2-methylphenylcarbamate) coated on silica particles (5 μm diameter) has been shown for the separation of cathinone derivatives designer drugs enantiomers by CEC [116]. The analysis of these drugs is very important because they exhibit a chemical structure quite similar to amphetamine with similar stimulating effects. They are frequently used and easily available even in internet. The method was optimized considering some experimental conditions such as mobile phase composition (buffer type and pH, organic additive), applied voltage and capillary temperature achieving the chiral separation of ten racemic cathinone derivatives. In addition a validation has also been performed. Later on in a comparative study utilizing four polysaccharide stationary phase in CEC, LC and SFC for the chiral resolution of ten cathinone- and amphetamine derivatives. CEC allowed the baseline separation of eight racemic couples of enantiomers. Best results were obtained employing amylose tris(3,5-dimethylphenylcarbamate) (ADH) column [117]. Polysaccharides have also been coated on silica core-shell particles for several applications in HPLC. It is known that these SPs can offer advantages over the porous ones, e.g., faster mass transfer, higher homogeneity with consequent shorter analysis time and higher efficiency. Coated silica core-shell particles have also been studied in CEC [118, 119] and the results compared with those obtained by nano-LC. In these studies good enantioresolution was obtained also observing a useful EOF that was ascribed to the presence of residual silanol groups on the SP and the higher density of the packed material into the column. Drugs such as warfarin, thalidomide, praziquantel and other standard enantiomers were resolved in their enantiomers by CEC utilizing silica coated (porous and core-shell particles) with cellulose derivative (10%, w/w) [118].

The content of polysaccharide coated onto the silica surface influences the EOF, higher concentration reduce this parameter covering the silanol groups. Bragg and Shamsi [120] proposed the use of a sulfonated CDMPC as a new chiral selector achieving the enantiomeric resolution of warfarin in less than two min with a column packed for seven cm.

A very interesting application related to the use of cellulose 3,5-di-(cholorophenylcarbamate) (CDCPC) has been reported by Chankvetadze et al. [121]. The CSP was used in both CEC and capillary liquid chromatography (CLC) for the enantiomers separation of norgestrel enantiomers studying some experimental parameters. Among them, the concentration of the coated polysaccharide strongly influenced the enantioresolution and efficiency. Using the lowest chiral selector concentration (5%-CDPC) number of theoretical plates about 145,000 were obtained by the electrodriven technique, while in CLC less than 45,000 were observed. The CEC method was applied to the enantiomeric purity of levonorgestrel pharmaceutical formulations.

High efficiency with N about 322,000 have been obtained utilizing different chiral selectors, e.g., strong chiral cation exchanger, based on penicillamine sulfonic acid, immobilized on thiol-modified silica particles (3.5 μm). Authors also investigated the reversion of ephedrine elution order that is of paramount importance when analyzing enantiomeric impurity using S- and R-CSP [122].

Silica containing bonded vancomycin has been successfully used for enantiomers separation of some drugs. Mirtazapine and two of its metabolites (8-hydroxymirtazapine and N-desmethylmirtazapine) were analyzed, after method optimization and validation in spiked human urine samples by CEC [123]. Enantiomers of studied compounds were baseline separated in less than 30 min. Vancomycin bonded to silica particles was also used for the analysis of venlafaxine and one of its metabolite (O-desmethylvenlafaxine) in plasma samples observing good LODs and relatively high efficiency (110,000 N/m, first enantiomer) [124]. In another study CEC was applied to the analysis of pharmaceutical formulations containing L-loxiglumide employing a column packed with silica modified with Hepta-Tyr antibiotic, a teicoplanin derivative [125].

An interesting application of CEC to the analysis of proteins has been reported by Zhang et al. [126]. Titanium oxide nanoparticles were coated on the capillary wall after polymerizing polydopamine. β-lactoglobulin and eight glycoisoforms of ovalbumin were separated. The optimized and validated OT-CEC method was applied to the analysis of acidic proteins in white egg. Egg white samples were also analyzed by Liu et al. [35] using OT-CEC in a different column where fibrous mesoporous silica nanoparticles were grafted on the capillary wall with a polymeric material (2-(dimethylamino)ethyl methacrylate). The column exhibited high reproducibility. A ligand exchange (LE) mechanism has been recently used for the separation of L- and D-malic acid in apple juice by OT-CEC. The capillary wall was modified by grafting polymerization of 3-chloro-2-hydroxypropyl methacrylate and then reacting with L-histidine and copper ions were added to the mobile phase [127]. Pressurized assisted CEC employing a monolithic column methacrylate based was successfully applied to the analysis of water and fat soluble vitamins present in tablets [128].

FUTURE TRENDS

In this chapter the main features of CEC have been presented. This analytical technique offers some advantages over conventional ones such as higher efficiency, lower consumption of both mobile phase and samples, lower costs and different separation mechanisms and easier coupling with MS. Therefore exhibits great potentiality and can be alternative or complementary to other techniques such as HPLC or electrophoresis. Although any SP studied in LC could be used, there is a fundamental requirement concerning the presence of charged/chargeable groups in order to generate a high EOF. CEC has been applied in various fields for the analysis of several compounds including drugs, enantiomers, proteins, nutraceuticals etc. Although the above summarized features, there is certainly a need for further studies related to the production of new stationary phases and construction of appropriate instrumentation.

CEC has been carried out mainly utilizing capillary column, however interesting results could be obtained utilizing chip technology or planar format. These two different modes could offer additional advantages, e.g., use of lower sample volumes and reduction/absence of dead volumes in the first case [129] and possibility to increase sample loading and two-dimensional separation in the last [130].

REFERENCES

[1] Pretorius V, Hopkins BJ, Schieke JD. (1974). Electro-osmosis: A new concept for high-speed liquid chromatography. *Journal of Chromatography A* 99:23-30.

[2] Knox JH, Grant IH. (1991) Electrochromatography in packed tubes using 1.5 to 50 ìm silica gels and ODS bonded silica gels. *Chromatographia* 32:317-28.

[3] Knox JH. (1988). Thermal effects and band spreading in capillary electro-separation. *Chromatographia* 26:329-37.

[4] Rathore AS, Horvath C. (1996) Separation parameters via virtual migration distances in high- performance liquid chromatography, capillary zone electrophoresis and electrokinetic chromatography. *Journal of Chromatography A* 743(2):231-46.

[5] Dermaux A, Sandra P. (1999). Applications of capillary electrochromatography. *Electrophoresis* 20:3027-65.

[6] Eeltink S, Rozing GP, Kok WT. (2003). Recent applications in capillary electrochromatography. *Electrophoresis* 24:3935-61.

[7] Huo Y, Kok WT. (2008). Recent applications in CEC. *Electrophoresis* 29:80-93.

[8] D'Orazio G, Asensio-Ramos M, Fanali C, Hernández-Borges J, Fanali S. (2016). Capillary electrochromatography in food analysis. *TrAC - Trends in Analytical Chemistry* 82:250-67.

[9] Rocco A, D'Orazio G, Aturki Z, Fanali S. Capillary Electrochromatography: A look at its features and potential in separation science. In: Fanali S, Haddad PR, Poole CF, Schoenmakers P, Lloyd D, editors. *Liquid Chromatography: Fundamental and Instrumentation.* Amsterdam: Elsevier BV; 2013. p. 469-92.

[10] Wiedmer SK, Riekkola M-L. Capillary electrochromatography. In: Fanali S, Haddad PR, Poole CF, Riekkola M-L, editors. *Liquid Chromatography*. Second ed. Amsterdam: Elsevier; 2017. p. 697-718.

[11] Rozing GP, Dermaux A, Sandra P. Instrumentation for capillary electrochromatography in Capillary Electrochromatography. In: Deyl Z, Svec F, editors. *Capillary electrochromatography.* Amsterdam: Elsevier Science B.V.; 2001. p. 39-85

[12] Aturki Z, Rocco A, Rocchi S, Fanali S. (2014). Current applications of miniaturized chromatographic and electrophoretic techniques in drug analysis. *Journal of Pharmaceutical and Biomedical Analysis* 101:194-220.

[13] Taylor MR, Teale P. (1997). Gradient capillary electrochromatography of drug mixtures with UV and electrospray ionisation mass spectrometric detection. *Journal of Chromatography A* 768(1):89-95.

[14] Huber CG, Choudhary G, Horvath C. (1997). Capillary electrochromatography with gradient elution. *Analytical Chemistry* 69(21):4429-36.

[15] Lister AS, Rimmer CA, Dorsey JG. (1998) Gradient elution electrochromatography with a flow-injection analysis interface. *Journal of Chromatography A* 828(1-2):105-12.

[16] Wang S, Jia L, Chen D. (2009) Pressurized CEC with gradient elution for separation of flavonoids from corn. *Journal of Separation Science* 32:388-93.

[17] Wu J-K, Yang C-S, Wu Y-S, Wang P-C, Tseng F-G. (2015). Continuous affinity-gradient nano-stationary phase served as a column for reversed-phase electrochromatography and matrix carrier in time-of-flight mass spectrometry for protein analysis. *Analytica Chimica Acta* 889:166-71.

[18] Tsuda T, Nomura K, Nakagawa G. (1982). Open-tubular microcapillary liquid chromatography with electro-osmosis flow using a UV detector. *Journal of Chromatography A* 248:241-7.

[19] Schurig V, Jung M, Mayer S, Fluck M, Segura S, Jakubetz H. (1995). Unified enantioselective capillary chromatography on a Chirasil-DEX stationary phase Advantages of column miniaturization. *Journal of Chromatography A* 694:119-28.

[20] Armstrong DW, Tang YB, Ward T, Nichols M. (1993) Derivatizedcyclodextrins immobilized on fused-silica capillaries for enantiomeric separations via capillary electrophoresis, gas chromatography, or supercritical fluid chromatography. *Analytical Chemistry* 65:1114-7.

[21] Francotte E, Jung M. (1996). Enantiomer separation by open tubular liquid chromatography and electrocromatography. *Chromatographia* 42:521-7.

[22] Bruggemann O, Freitag R, Whitcombe MJ, Vulfson EN. (1997). Comparison of polymer coatings of capillaries for capillary

electrophoresis with respect to their applicability to molecular imprinting and electrochromatography. *Journal of Chromatography A* 781(1-2):43-53.

[23] Tan ZJ, Remcho VT. (1998). Bonded polymethacrylate stationary phases for open tubular liquid chromatographic and electrochromatographic separations. *Journal of Microcolumn Separations* 10(1):99-105.

[24] Liu Z, Zou HF, Ni JY, Zhang YK. (1999). Open tubular capillary electrochromatography with adsorbed stationary phase. *Analytica Chimica Acta* 378(1-3):73-6.

[25] Guan N, Zeng ZR, Wang YC, Fu EQ, Cheng JK. (2000). Open tubular capillary electrochromatography in fused-silica capillaries chemically bonded with macrocyclic dioxopolyamine. *Analytica Chimica Acta* 418(2):145-51.

[26] Liu Z, Otsuka K, Terabe S. (2001). Chiral separation by open tubular capillary electrochromatography with adsorbed avidin as a stationary phase. *Journal of Separation Science* 24(1):17-26.

[27] Matyska MT, Pesek JJ, Boysen RI, Hearn MTW. (2001). Characterization of open tubular capillary electrochromatography columns for the analysis of synthetic peptides using isocratic conditions. *Analytical Chemistry* 73(21):5116-25.

[28] Li HB, Zeng ZR, Xie CH, Chen YY. (2002). Preparation and application of a novel type of calix[6]crown coated capillary for open-tubular capillary electrochromatography. *Chromatographia* 55(9-10):591-4.

[29] Dawson GB, Matyska MT, Pesek JJ, Seipert RR. (2004). Electrochromatographic studies of etched capillaries modified with a cyanopentoxy biphenyl liquid crystal. *Journal of Chromatography A* 1047(2):299-303.

[30] Yang L, Guihen E, Holmes JD, Loughran M, O'Sullivan GP, Glennon JD. (2005) Gold nanoparticle-modified etched capillaries for open-tubular capillary electrochromatography. *Analytical Chemistry* 77(6):1840-6.

[31] Xu L, Sun Y. (2008). Protein separation by open tubular capillary electrochromatography employing a capillary coated with phenylalanine functionalized tentacle-type polymer under both cathodic and anodic electroosmotic flows. *Journal of Chromatography A* 1183(1-2):129-34.

[32] Li M, Liu X, Jiang F, Guo L, Yang L. (2011). Enantioselective open-tubular capillary electrochromatography using cyclodextrin-modified gold nanoparticles as stationary phase. *Journal of Chromatography A* 1218(23):3725-9.

[33] Liu X, Liu X, Li M, Guo L, Yang L. (2013). Application of graphene as the stationary phase for open-tubular capillary electrochromatography. *Journal of Chromatography A* 1277:93-7.

[34] Fang LL, Wang P, Wen XL, Guo X, Luo LD, Yu J, et al. (2017). Layer-by-layer self-assembly of gold nanoparticles/thiols β-cyclodextrin coating as the stationary phase for enhanced chiral differentiation in open tubular capillary electrochromatography. *Talanta* 167:158-65.

[35] Liu Y, Liu Q, Yu H, Sun S, Xue Y, Wang Y, et al. (2017). Polymer-modified fibrous mesoporous silica nanoparticles as coating material for open-tubular capillary electrochromatography. *Journal of Chromatography A* 1499:196-202.

[36] Rocchi S, Fanali C, Fanali S. (2015). Use of a Novel Sub-2 μm Silica Hydride Vancomycin Stationary Phase in Nano-Liquid Chromatography. II. Separation of Derivatized Amino Acid Enantiomers. *Chirality* 27(11):767-72.

[37] Rocchi S, Rocco A, Pesek JJ, Matyska MT, Capitani D, Fanali S. (2015). Enantiomers separation by nano-liquid chromatography: Use of a novel sub-2μm vancomycin silica hydride stationary phase. *Journal of Chromatography A* 1381:149-59.

[38] Fanali S, Rocchi S, Chankvetadze B. (2013). Use of novel phenyl-hexyl core-shell particles in nano-LC. *Electrophoresis* 34:1737-42.

[39] Robson MM, Cikalo MG, Myers P, Euerby MR, Bartle KD. (1997). Capillary electrochromatography - a review. *Journal of Microcolumn Separations* 9(5):357-72.

[40] Poole CF. *The essence of Chromatography*. Amsterdam: Elsevier Science B.V., 2003.

[41] Choudhary G, Horvath C, Banks JF. (1998). Capillary electrochromatography of biomolecules with on-line electrospray ionization and time-of-flight mass spectrometry. *Journal of Chromatography A* 828(1-2):469-80.

[42] Zhang B, Liu Q, Yang L, Wang Q. (2013). Performance of single particle fritted capillary columns in electrochromatography. *Journal of Chromatography A* 1272:136-40.

[43] Chen JR, Dulay MT, Zare RN, Svec F, Peters E. (2000). Macroporous photopolymer frits for capillary electrochromatography. *Analytical Chemistry* 72(6):1224-7.

[44] Rocco A, Fanali S. (2008). Capillary electrochromatography without external pressure assistance. Use of packed columns with a monolithic inlet frit. *Journal of Chromatography A* 1191:263-7.

[45] D'Orazio G, Fanali S. (2012). C18 silica packed capillary columns with monolithic frits prepared with UV light emitting diode: Usefulness in nano-liquid chromatography and capillary electrochromatography. *Journal of Chromatography A* 1232:176-82.

[46] Huber CG, Holzl G. Hyphenation of capillary electrochromatography and mass spectrometry: instrumental aspects, separation systems and applications. In: Deyl Z, Svec F, editors. *Capillary electrochromatography*. Amsterdam: Elsevier; 2001. p. 271-316.

[47] Tanaka N, Kobayashi H, Ishizuka N, Minakuchi H, Nakanishi K, Hosoya K, et al. (2002). Monolithic silica columns for high-efficiency chromatographic separations. *Journal of Chromatography A* 965:35-49.

[48] Ou J, Li X, Feng S, Dong J, Dong X, Komg L, et al. (2007). Preparation and Evaluation of a Molecularly Imprinted Polymer Derivatized Silica Monolithic Column for Capillary Electrochromatography and Capillary Liquid Chromatography. *Analytical Chemistry* 79:639-46.

[49] Li Liao J, Chen N, Ericson C, Hjerten S. (1996) Preparation of continuous beds derivatized with one-step alkyl and sulfonate groups

for capillary electrochromatography. *Analytical Chemistry* 68:3468-72.

[50] Tang QL, Lee ML. (2000). Continuous-bed columns containing sol-gel bonded packing materials for capillary electrochromatography. *HRC-Journal of High Resolution Chromatography* 23(1):73-80.

[51] Ganewatta N, El Rassi Z. (2018). Organic polymer-based monolithic stationary phases with incorporated nanostructured materials for HPLC and CEC. *Electrophoresis* 39(1):53-66.

[52] Mikšík I. (2017). Capillary electrochromatography of proteins and peptides (2006–2015). *Journal of Separation Science* 40(1):251-71.

[53] Wei ZH, Mu LN, Huang YP, Liu ZS. (2017). Imprinted monoliths: Recent significant progress in analysis field. *TrAC - Trends in Analytical Chemistry* 86:84-92.

[54] Guo J, Lin Y, Xiao Y, Crommen J, Jiang Z. (2016). Recent developments in cyclodextrin functionalized monolithic columns for the enantioseparation of chiral drugs. *Journal of Pharmaceutical and Biomedical Analysis* 130:110-25.

[55] Yang S, Wang Y, Jiang Y, Li S, Liu W. (2016). Molecularly imprinted polymers for the identification and separation of chiral drugs and biomolecules. *Polymers* 8(6).

[56] Healey R, Ghanem A. (2013). An insight to chiral monolith for enantioselective nano and micro HPLC: Preparation and applications. *Chirality* 25(6):314-23.

[57] Adly FG, Antwi NY, Ghanem A. (2016). Cyclodextrin-Functionalized Monolithic Capillary Columns: Preparation and Chiral Applications. *Chirality* 28(2):97-109.

[58] Ganewatta N, El Rassi Z. (2018). Poly(glyceryl monomethacrylate-co-ethylene glycol dimethacrylate) monolithic columns with incorporated bare and surface modified gluconamide fumed silica nanoparticles for hydrophilic interaction capillary electrochromatography. *Talanta* 179:632-40.

[59] Li XX, Zhang LS, Wang C, Huang YP, Liu ZS. (2018). Green synthesis of monolithic column incorporated with graphene oxide

using room temperature ionic liquid and eutectic solvents for capillary electrochromatography. *Talanta* 178:763-71.

[60] Liu C, Deng Q, Fang G, Dang M, Wang S. (2017). Capillary electrochromatography immunoassay for alpha-fetoprotein based on poly(guanidinium ionic liquid) monolithic material. *Analytical Biochemistry* 530:50-6.

[61] Mao Z, Chen Z. (2017). Monolithic column modified with bifunctional ionic liquid and styrene stationary phases for capillary electrochromatography. *Journal of Chromatography A* 1480:99-105.

[62] Zhang LS, Zhao QL, Li XX, Li XX, Huang YP, Liu ZS. (2016). Green synthesis of mesoporous molecular sieve incorporated monoliths using room temperature ionic liquid and deep eutectic solvents. *Talanta* 161:660-7.

[63] Hong T, Chen X, Xu Y, Cui X, Bai R, Jin C, et al. (2016). Preparation of graphene oxide-modified affinity capillary monoliths based on three types of amino donor for chiral separation and proteolysis. *Journal of Chromatography A* 1456:249-56.

[64] Zhao H, Wang Y, Cheng H, Shen Y. (2016). Graphene oxide decorated monolithic column as stationary phase for capillary electrochromatography. *Journal of Chromatography A* 1452:27-35.

[65] Ganewatta N, El Rassi Z. (2018). Monolithic capillary columns consisting of poly(glycidyl methacrylate-: Co-ethylene glycol dimethacrylate) and their diol derivatives with incorporated hydroxyl functionalized multiwalled carbon nanotubes for reversed-phase capillary electrochromatography. *Analyst* 143(1):270-9.

[66] Zhao H, Wang Y, Cheng H, Wang Y. (2017). Fabrication of single-walled carbon nanohorns incorporated a monolithic column for capillary electrochromatography. *Journal of Separation Science* 40(16):3343-50.

[67] Miao C, Bai R, Xu S, Hong T, Ji Y. (2017). Carboxylated single-walled carbon nanotube-functionalized chiral polymer monoliths for affinity capillary electrochromatography. *Journal of Chromatography A* 1487:227-34.

[68] Navarro-Pascual-Ahuir M, Lucena R, Cárdenas S, Ramis-Ramos G, Valcárcel M, Herrero-Martínez JM. (2014). UV-polymerized butyl methacrylate monoliths with embedded carboxylic single-walled carbon nanotubes for CEC applications. *Analytical and Bioanalytical Chemistry* 406(25):6329-36.

[69] Aşır S, Sarı D, Derazshamshir A, Yılmaz F, Şarkaya K, Denizli A. (2017). Dopamine-imprinted monolithic column for capillary electrochromatography. *Electrophoresis* 38(22-23):3003-12.

[70] Al-Massaedh AA, Pyell U. (2017). Mixed-Mode Acrylamide-Based Continuous Beds Bearing tert-Butyl Groups for Capillary Electrochromatography Synthesized Via Complexation of N-tert-Butylacrylamide with a Water-Soluble Cyclodextrin. Part II: Effect of Capillary Size and Polymerization Conditions on Morphology and Chromatographic Efficiency. *Chromatographia* 80(11):1669-82.

[71] Xiong L, Li R, Ji Y. (2017). Preparation and performance characterization of gold nanoparticles modified chiral capillary electrochromatography stationary phase. *Chinese Journal of Chromatography* (Se Pu) 35(7):712-8.

[72] Xu S, Mo R, Jin C, Cui X, Bai R, Ji Y. (2017). Mesoporous silica nanoparticles incorporated hybrid monolithic stationary phase immobilized with pepsin for enantioseparation by capillary electrochromatography. *Journal of Pharmaceutical and Biomedical Analysis* 140:190-8.

[73] Hong T, Zheng Y, Hu W, Ji Y. (2014). Preparation and evaluation of bovine serum albumin immobilized chiral monolithic column for affinity capillary electrochromatography. *Analytical Biochemistry* 464:43-50.

[74] Hong T, Chi C, Ji Y. (2014). Pepsin-modified chiral monolithic column for affinity capillary electrochromatography. *Journal of Separation Science* 37(22):3377-83.

[75] Zhang LS, Du PY, Gu W, Zhao QL, Huang YP, Liu ZS. (2016). Monolithic column incorporated with lanthanide metal-organic framework for capillary electrochromatography. *Journal of Chromatography A* 1461:171-8.

[76] Qiu D, Li F, Zhang M, Kang J. (2016). Preparation of phosphorylcholine-based hydrophilic monolithic column and application for analysis of drug-related impurities with capillary electrochromatography. *Electrophoresis* 37(12):1725-32.

[77] Dixit S, Park JH. (2015). Enantioseparation of basic chiral drugs on a carbamoylated erythromycin-zirconia hybrid monolith using capillary electrochromatography. *Journal of Chromatography A* 1416:129-36.

[78] Carrasco-Correa EJ, Ramis-Ramos G, Herrero-Martínez JM. (2013). Methacrylate monolithic columns functionalized with epinephrine for capillary electrochromatography applications. *Journal of Chromatography A* 1298:61-7.

[79] Dixit S, Lee IS, Park JH. (2017). Carbamoylated azithromycin incorporated zirconia hybrid monolith for enantioseparation of acidic chiral drugs using non-aqueous capillary electrochromatography. *Journal of Chromatography A* 1507:132-40.

[80] Tran LN, Park JH. (2015). Enantiomer separation of acidic chiral compounds on a quinine-silica/zirconia hybrid monolith by capillary electrochromatography. *Journal of Chromatography A* 1396:140-7.

[81] Park JM, Park JH. (2014). Enantiomer separations of basic chiral compounds by capillary electrochromatography on a phosphated β-cyclodextrin-modified zirconia monolith. *Journal of Chromatography A* 1339:229-33.

[82] Wall W, Li J, El Rassi Z. (2002). Electrically driven micro separation methods for pesticides and metabolites Part VII: capillary electrophoresis and electrochromatography of derivatized and underivatized phenol pesticidic metabolites. Preconcentration and laser induced fluorescence detection of dilute samples. *Journal of Separation Science* 25:1231-44.

[83] Kuban P, Kuban P, Kuban V, Hauser PC, Bocek P. (2008). Capillary electrochromatography of inorganic cations in open tubular columns with a controllable capacity multilayered stationary phase architecture. *Journal of Chromatography A* 1190:377-82.

[84] Wu W, Yuan X, Wu X, Lin X, Xie Z. (2010). Analysis of phenolic xenoestrogens by pressurized CEC with amperometric detection. *Electrophoresis* 31:1011-8.

[85] Bragg W, Norton D, Shamsi SA. (2008). Optimized separation of β-blockers with multiple chiral centers using capillary electrochromatography-mass spectrometry. *Journal of Chromatography B* 875(1):304-16.

[86] Cheng YJ, Huang SH, Singco B, Huang HY. (2011). Analyses of sulfonamide antibiotics in meat samples by on-line concentration capillary electrochromatography-mass spectrometry. *Journal of Chromatography A* 1218(42):7640-7.

[87] Warriner RN, Craze AS, Games DE, Lane SJ. (1998). Capillary electrochromatography mass spectrometry - a comparison of the sensitivity of nanospray and microspray ionization techniques. *Rapid Communications in Mass Spectrometry* 12(17):1143-9.

[88] Schmeer K, Behnke B, Bayer E. (1995). Capillary electrochromatography-electrospray mass spectrometry: A microanalysis technique. *Analytical Chemistry* 67(20):3656-8.

[89] D'Orazio G, Fanali S. (2010). Coupling capillary electrochromatography with mass spectrometry by using a liquid-junction nano-spray interface. *Journal of Chromatography A* 1217(25):4079-86.

[90] Koerner T, Turck K, Brown L, Oleschuk RD. (2004). Porous Polymer Monolith Assisted Electrospray. *Analytical Chemistry* 76:6456-60.

[91] Desiderio C, Fanali S. (2000). Capillary electrochromatography and capillary electrochromatography–electrospray mass spectrometry for the separation of non-steroidal anti-inflammatory drugs. *Journal of Chromatography A* 895:123-32.

[92] Foret F, Zhou H, Gangi E, Karger BL. (2000). Subatmospheric electrospray interface for coupling of microcolumn separations with mass spectrometry. *Electrophoresis* 21:1363-71.

[93] Fanali S, D'Orazio G, Foret F, Kleparnik K, Aturki Z. (2006). On-line CE-MS using pressurized liquid-junction nanoflow electrospray interface and surface-coating capillaries. *Electrophoresis* 27:4666-73.

[94] D'Orazio G, Fanali S. (2013). Pressurized nano-liquid-junction interface for coupling capillary electrochromatography and nano-liquid chromatography with mass spectrometry. *Journal of Chromatography A* 1317:67-76.

[95] Van Hoenacker G, Bosch Vd, Rozing G, Sandra P. (2001). Recent applications of capillary electrochromatography. *Electrophoresis* 22:4064-103.

[96] Eeltink S, Kok WT. (2006). Recent applications in capillary electrochromatography. *Electrophoresis* 27:84-96.

[97] Declerck S, Vander Heyden Y, Mangelings D. (2016). Enantioseparations of pharmaceuticals with capillary electrochromatography: A review. *Journal of Pharmaceutical and Biomedical Analysis* 130:81-99.

[98] Carrasco-Correa EJ, Simo-Alfonso EF, Ramis-Ramos G, Herrero-Martinez JM. (2017) Application of organic monolithic materials to enantioseparation in capillary separation techniques. *Current Medicinal Chemistry* 24(8):781-95.

[99] D'Orazio G, Fanali C, Asensio-Ramos M, Fanali S. (2017) Chiral separations in food analysis. TrAC - *Trends in Analytical Chemistry* 96:151-71.

[100] Guihen E. 2017 Recent highlights in electro-driven separations-selected applications of alkylthiol gold nanoparticles in capillary electrophoresis and capillary electrochromatography. *Electrophoresis* 38(17):2184-92.

[101] Greño M, Castro-Puyana M, García MÁ, Marina ML. (2018) Analysis of antibiotics by CE and CEC and their use as chiral selectors: An update. *Electrophoresis* 39(1):235-59.

[102] Rutkowska M, Płotka-Wasylka J, Morrison C, Wieczorek PP, Namieśnik J, Marć M. (2018). Application of molecularly imprinted polymers in analytical chiral separations and analysis. *TrAC - Trends in Analytical Chemistry* 102:91-102.

[103] Orlandini S, Furlanetto S, Pinzauti S, D'Orazio G, Fanali S. (2004). Analysis of ketorolac and its related impurities by capillary electrochromatography. *Journal of Chromatography A* 1044:295-303.

[104] Pucci V, Mandrioli R, Raggi MA, Fanali S. (2004). Reversed-phase capillary electrochromatography for the simultaneous etermination of acetylsalicylic acid, paracetamol, and caffeine in analgesic tablets. *Electrophoresis* 25:615-21.

[105] Aturki Z, D'Orazio G, Fanali S. (2005). Rapid assay of vitamin E in vegetable oils by reversed-phase capillary electrochromatography. *Electrophoresis* 26:798–803.

[106] Aturki Z, Fanali S, D'Orazio G, Rocco A, Rosati C. (2008). Analysis of phenolic compounds in extra virgin olive oil by using reversed-phase capillary electrochromatography. *Electrophoresis* 29:1643-50.

[107] Uysal UD, Aturki Z, Raggi MA, Fanali S. (2009). Separation of catechins and methylxanthines in tea samples by capillary electrochromatography. *Journal of Separation Science* 32:1002-10.

[108] Hernández-Mesa M, D'Orazio G, Rocco A, García-Campaña AM, Blanco CC, Fanali S. (2015). Capillary electrochromatography-mass spectrometry for the determination of 5-nitroimidazole antibiotics in urine samples. *Electrophoresis* 36(20):2606-15.

[109] Musenga A, Mandrioli R, Ferranti A, D'Orazio G, Fanali S, Raggi MA. (2007). Analysis of aromatic and terpenic constituents of pepper extracts by capillary electrochromatography. *Journal of Separation Science* 30:612-9.

[110] Abidi SL. (2004). Capillary electrochromatography of sterols and related steryl esters derived from vegetable oils. *Journal of Chromatography A* 1059:199-208.

[111] Ye F, Xie Z, Wu X, Lin X. (2006) Determination of pyrethroid pesticide residues in vegetables by pressurized capillary electrochromatography. *Talanta* 69:97-102.

[112] Yang FQ, Li S, Li P, Wang YT. (2007). Optimization of CEC for simultaneous determination of eleven nucleosides and nucleobases in Cordyceps using central composite design. *Electrophoresis* 28:1681–8.

[113] Chen XJ, Zhao J, Meng Q, Li SP, Wang YT. (2009). Simultaneous determination of five flavonoids in licorice using pressurized liquid extraction and capillary electrochromatography coupled with peak

suppression diode array detection. *Journal of Chromatography A* 1216:7329-35.

[114] Zhang B, Cai X, Yin Y, Li X, Lu H, Wu X. (2017). Analysis of biogenic amines in foods by capillary electrochromatography coupled with laser induced fluorescence detection. *Chinese Journal Chromatography* (Se Pu) 35:344-50.

[115] Sander LC, Pursch M, Marker B, Wise SA. (1999). Separation of carotenoid isomers by capillary electrochromatography with c-30 stationary phases. *Analytical Chemistry* 71(16):3477-83.

[116] Aturki Z, Schmid MG, Chankvetadze B, Fanali S. (2014). Enantiomeric separation of new cathinone derivatives designer drugs by capillary electrochromatography using a chiral stationary phase, based on amylose tris(5-chloro-2-methylphenylcarbamate). *Electrophoresis* 35:3242-9.

[117] Albals D, VanderHeyden Y, Schmid MG, Chankvetadze B, Mangelings D. (2016). Chiral separations of cathinone and amphetamine-derivatives: Comparative study between capillary electrochromatography, supercritical fluid chromatography and three liquid chromatographic modes. *Journal of Pharmaceutical and Biomedical Analysis* 121:232-43.

[118] Fanali S, D'Orazio G, Farkas T, Chankvetadze B. (2012). Comparative performance of capillary columns made with totally porous and core-shell particles coated with a polysaccharide-based chiral selector in nano-liquid chromatography and capillary electrochromatography. *Journal of Chromatography A* 1269:136-42.

[119] Rocchi S, Fanali S, Farkas T, Chankvetadze B. (2014). Effect of content of chiral selector and pore size of core–shell type silica support on the performance of amylose tris(3,5-dimethylphenyl-carbamate)-based chiral stationary phases in nano-liquid chromatography and capillary electrochromatography. *Journal of Chromatography A* 1363:363-71.

[120] Bragg W, Shamsi SA. (2013). High Throughput Analysis of Chiral Compounds Using Capillary Electrochromatography (CEC) and

CEC-Mass Spectrometry with Cellulose Based Stationary Phases. *Separation Science and Technology (Philadelphia)* 48(17):2589-99.

[121] Chankvetadze B, Kartozia I, Yamamoto C, Okamoto Y, Balschke G. (2003). Comparative study on the application of capillary liquid chromatography and capillary electrochromatography for investigation of enantiomeric purity of the contraceptive drug levonorgestrel. *Journal of Pharmaceutical and Biomedical Analysis* 30:1897-906.

[122] Bicker W, Hebenstreit D, Lammerhofer M, Lindner W. (2003). Enantiomeric impurity profiling in ephedrine samples by enantioselective capillary electrochromatography. *Electrophoresis* 24:2532-42.

[123] Aturki Z, Scotti V, D'Orazio G, Rocco A, Raggi MA, Fanali S. (2007). Enantioselective separation of the novel antidepressant mirtazapine and its main metabolites by CEC. *Electrophoresis* 28(15):2717-25.

[124] Fanali S, Rudaz S, Veuthey JL, Desiderio C. (2001). Use of vancomycin silica stationary phase in packed capillary electrchromatography. II: Enantiomer separation of venlafaxine and O-desmethylvenlafaxine in human plasma. *Journal of Chromatography A* 919:195-203.

[125] Fanali S, D'Orazio G, Quaglia MG, Rocco A. (2004). Use of a Hepta-Tyr antibiotic modified silica stationary phase for the enantiomeric resolution of d,l-loxiglumide by electrochromatography and nano-liquid chromatography. *Journal of Chromatography A* 1051:247-52.

[126] Zhang Y, Wang W, Ma X, Jia L. (2016). Polydopamine assisted fabrication of titanium oxide nanoparticles modified column for proteins separation by capillary electrochromatography. *Analytical Biochemistry* 512:103-9.

[127] Aydoğan C, Karakoç V, Denizli A. (2015). Chiral ligand-exchange separation and determination of malic acid enantiomers in apple juice by open-tubular capillary electrochromatography. *Food Chemistry* 187:130-4.

[128] Yamada H, Kitagawa S, Ohtani H. (2013). Simultaneous separation of water- and fat-soluble vitamins in isocratic pressure-assisted

capillary electrochromatography using a methacrylate-based monolithic column. *Journal of Separation Science* 36(12):1980-5.

[129] Masár M, Sydes D, Luc M, Kaniansky D, Kuss H-M. (2009). Determination of ammonium, calcium, magnesium, potassium and sodium in drinking waters by capillary zone electrophoresis on a column coupling chip. *Journal of Chromatography A* 1216:6252-5.

[130] Dzido TH, Plocharz PW, Chomicki A, Halka-Grysinska A, Polak B. (2011). Pressurized planar electrochromatography. *Journal of Chromatography A* 1218:2636-47.

In: Capillary Electrophoresis … ISBN: 978-1-53615-223-4
Editors: O. Núñez and X. Subirats

Chapter 6

MICELLAR AND MICROEMULSION ELECTROKINETIC CAPILLARY CHROMATOGRAPHY AS SURROGATE MODELS FOR LIPOPHILICITY AND BIOLOGICAL ACTIVITY

Xavier Subirats[1,*] and Oscar Núñez[1,2,3]
[1]Department of Chemical Engineering and Analytical Chemistry
University of Barcelona, Barcelona, Spain
[2]Research Institute in Food Nutrition and Food Safety,
University of Barcelona, Barcelona, Spain
[3]Serra Húnter Fellow, Generalitat de Catalunya, Spain

ABSTRACT

In this chapter chromatographic methods are examined for lipophilicity determination and as surrogate models for biological activity, particularly micellar and microemulsion electrokinetic chromatography

* Corresponding Author's Email: xavier.subirats@ub.edu.

(MEKC and MEEKC). The fundamentals of these techniques are presented, together with the relationships of lipophilicity or biological activity with the mass distribution ratios (or capacity factors) obtained from the migration of analytes in the capillary electrophoretic system. The Abraham's solvation model, based on linear free energy relationships, is presented as a very powerful tool for the characterization of partition processes and their comparison in order to find the best possible matches between chromatographic, physicochemical and biological systems.

Keywords: biological activity, lipophilicity, MEKC, MEEKC

ABBREVIATIONS FOR SURFACTANTS

ALE	sodium *N*-lauroyl-*N*-methyl-*β*-alaninate,
AOT	bis(2-ethylhexyl)sodium sulfosuccinate,
Brij 35	polyethyleneglycol lauryl ether
DHAB	dihexadecyl-trimethylammonium bromide,
HTAB	hexadecyltrimethylammonium bromide,
LMT	sodium *N*-lauroyl-*N*-methyltaurate,
LPFOS	lithium perfluorooctane sulfonate,
SC	sodium cholate,
SDCV	sodium dodecyl carbonyl valine,
SDS	sodium dodecyl sulfate,
SLN	sodium *N*-lauroylsarcosinate,
SLSA	sodium dodecyl sulfoacetate,
STC	sodium taurocholate,
STDC	sodium taurodeoxy-cholate,
TTAB	tetradecyltrimethylammonium bromide.

INTRODUCTION

Lipophilicity is a fundamental property affecting the absorption, distribution, metabolism, excretion, and toxicity of chemical compounds in biological systems [1]. It measures the capacity of a molecule to go through cell membranes by passive diffusion and it is often correlated to biological activity [2]. Thus, lipophilicity measurement is clearly relevant for pharmaceutical companies in the discovery process of potential drug candidates, but it also affects different chemical industries due to regulatory safety assessments. For instance, according to the European REACH Regulation (EC No 1907/2006), lipophilicity must be reported for any organic compound produced in quantities of one tonne or more per year.

Lipophilicity is traditionally assessed by the partition ratio (or partition coefficient) of the studied compound between to immiscible solvents, 1-octanol and water (or aqueous buffer), defined as:

$$\log P_{\text{o/w}} = \log_{10} \frac{[\text{compound}]_{\text{organic}}}{[\text{compound}]_{\text{aqueous}}} \tag{1}$$

In case of solutes with acid/base properties, generally log $P_{\text{o/w}}$ is referred to the neutral species, since the partition of the salt depends on the nature of the counterion and it is about 3-4 units lower [3]. Octanol was choosen as a simple surrogate model of lipids allowing interlaboratory comparison, showing the alcohol molecule dipolarity characteristics, with a non-polar hydrocarbon substructure and a hydrogen-bonding hydroxyl moiety.

The reference method for log $P_{\text{o/w}}$ determination is *shake-flask* [4–6], based on the solute equilibration between octanol and aqueous buffer for several hours (up to 24-48h), and followed by the determination of the concentration in one or both phases often by liquid chromatography. However, this method requires samples of high purity and it is very time consuming, which may be an inconvenience in early phases of the drug discovery process when a high throughput is required. In order to overcome (at least partially) these limitations other experimental methods can be employed for log $P_{\text{o/w}}$ measurement provided that they show an acceptable level of quality assurance [7]. This is the case of the potentiometric titration procedure in octanol-water mixtures [8, 9], which reduces considerably the required experimental time to a few hours obtaining equivalent results to shake-flask method, but this procedure can only be applied to compounds with acid/base properties, sharing in common with shake-flask the need of high purity samples [10]. Therefore, with the aim of overcoming sample purity requirements and the reduction of analysis time to minutes, chromatographic approaches appeared to be a very interesting alternative to shake-flask and potentiometry for lipophilicity determination.

CHROMATOGRAPHIC METHODS FOR LIPOPHILICITY ASSESSMENT

Relationships between Partition Ratios in Different Solvents

In 1950s Collander found linear relationships between partition ratios obtained in different alcohol-water solvents ($\log P_1 = a \log P_2 + b$), that were extended in the following years by different authors to several organic solvents such cyclohexane, heptane, carbon tetrachloride, xylene, toluene, benzene, chloroform, carbon tetrachloride, isopentyl acetate, ether… Some compounds with hydrogen-bonding capabilities presented some deviations from this linear relationship depending on the hydrogen-bonding ability of the solvent. For instance, correlations for phenols between octanol and cyclohexane are generally poor; there are hydrogen-bonding interactions between phenols (donor) and octanol (acceptor), whereas this reciprocal action can not take place with the aprotic cyclohexane. However, with the introduction of a correction for hydrogen-bonding acidity of phenols these correlations are significantly improved [11].

Conventional Reversed-Phase Liquid Chromatography

Liquid-liquid chromatographic techniques are based on the different partition of analytes between a stationary phase confined in a column and a mobile phase flowing along. In reversed-phase mode typical stationary phases consist of a silica-base suport functionalized with a non-polar molecule (e.g., C18 or octadecylsilane), and mobile phases are made of buffered hydroorganic mixtures containing generally acetonitrile or methanol as organic modifiers. Thus, the chromatographic retention factor (k), which represents the ratio of the amounts of a solute in the stationary and mobile phases, can be defined as:

$$k = \frac{C_{SP} V_{SP}}{C_{MP} V_{MP}} = K \frac{V_{SP}}{V_{MP}} \quad (2)$$

where C_{SP} and C_{MP} refer to solute concentrations in the stationary and mobile phases, respectively; V_{SP} and V_{MP} are the corresponding volumes of both phases inside the column, and K is the equilibrium constant for the partition process. Notice that in normal chromatographic practice, the retention factor is easily determined from the retention times of the analyte and a non-retained marker ($k = (t_R - t_0)/t_0$). As long as the phase ratio (V_{SP}/V_{MP}) remains constant, which is mainly true for a reversed-phase column, retention factors can be directly related with partition constants, leading to linear relationships between retention factors (log k) and partition ratios (log K or log P). This approach have been particularly assayed for lipophilicity assessment, trying to correlate chromatographic retention in reversed-phase with log $P_{o/w}$ values (log $P_{o/w} = a \log k + b$), following a similar approach to that described by Collander (for instance, reference [12] and cited therein). However, as comented before for phenolic compounds and cyclohexane/octanol as solvents, chromatographic C18 is an alkane lacking hydrogen-bond capabilities, whereas octanol shows hydrogen-bond basicity. Consequently, linear correlations between log k and log $P_{o/w}$ need to be improved by molecular descriptors accounting for solute hydrogen-bonding acidity [13, 14]. This descriptors can be as simple as the number of hydrogen-bond donors of a molecule, but more accurate log $P_{o/w}$ values from chromatographic measurements are obtained when more complex descriptors are considered, either experimentally measured or calculated from their 2D or 3D structure.

Micellar and Microemulsion Electrokinetic Chromatography

Electrokinetic separations using open-tubular column capillaries and micellar solutions as pseudostationary phases were firstly reported by Terabe in 1984 [15]. This separation technique, known as micellar electrokinetic chromatography (MEKC), is based on the partition of analytes between the micelles formed from a surfactant and an aqueous buffer filling the capillary. Under the influence of an electric field, charged micelles migrate towards the corresponding electrode depending on their charge (positive or negative), and a window is established between the migration of a neutral analyte only under the influence of the electroosmotic flow and the charged micelles migrating as a narrow band. The analyzed solutes, which are uncharged at the pH of the running buffer, migrate within the elution window depending on their partition between micelles and bulk aqueous buffer. The higher the partition into the micelles, the closer the migration of the neutral analyte to that of the micelles. In MEKC the capacity factor (or the mass distribution ratio) of the solute between micelles and aqueous buffer, is given by:

$$k = \frac{t_{a} - t_{eo}}{t_{eo}(1 - t_{a}/t_{mc})} \tag{3}$$

where t_{eo}, t_a, and t_{mc} are the migration times of a neutral analyte with negligible partition into the micelle (e.g., DMSO, as electroosmotic flow marker), the analyte, and a very lipophilic solute with negligible partition into the aqueous buffer (e.g., dodecanophenone, as micelle marker), respectively. The separation principle of MEKC in fused silica capillaries using sodium dodecyl sulfate (SDS) as surfactant is schematically shown in Figure 1. Notice that the term $(1\text{-}t_a/t_{mc})$ in Equation (3) would be 1 in case of micelles being immobilized inside the capillary (infinite t_{mc}), leading to the conventional expression for the measurement of retention factors in column chromatography.

Contrarily to conventional chromatography where stationary phase is packed inside a column, in MEKC the properties of the pseudo-stationary phase can be easily modified using surfactants of different nature (anionic, cationic, zwitterionic non-ionic, and bile salts) or a mixture of them. For instance, bile salt surfactants were reported to be more appropriate than SDS or cetyltrimethylammonium bromide (CTAB) micellar systems for the estimation of log $P_{o/w}$ values from log k measurements, because of their better similarity concerning hydrogen-bonding interactions with octanol [16].

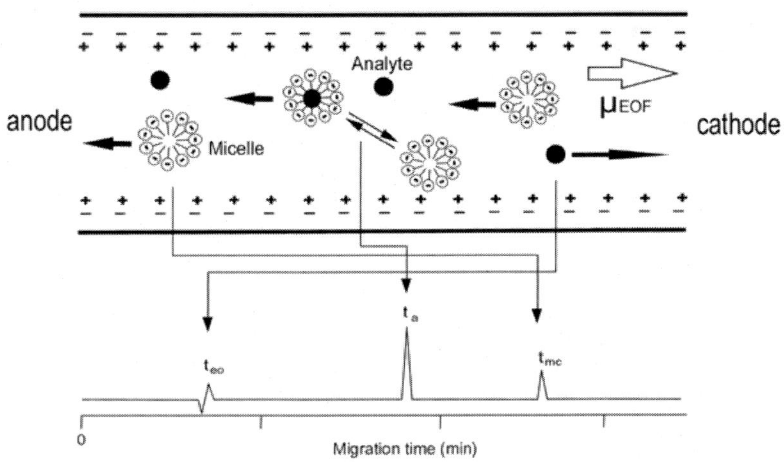

Figure 1. Schematic representation of the separation mechanism of MEKC using anionic micelles (e.g., SDS) and positive polarity. From ref. [17], with permission.

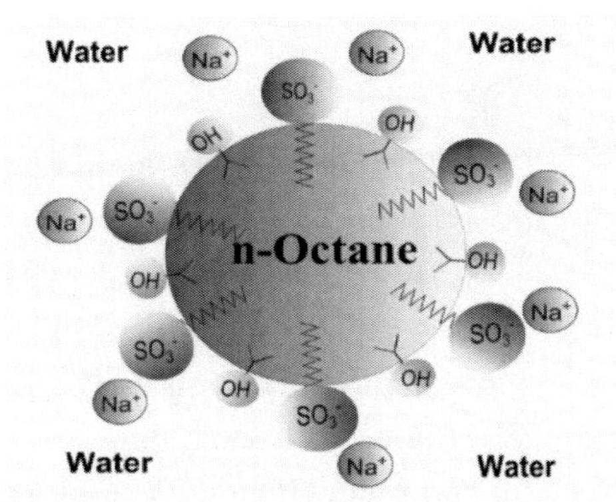

Figure 2. Schematic representation of an oil-in-water microemulsion droplet consisting of octane as oil phase, SDS as surfactant and isopropanol as co-surfactant. Adapted from ref. [22] (CC BY 4.0).

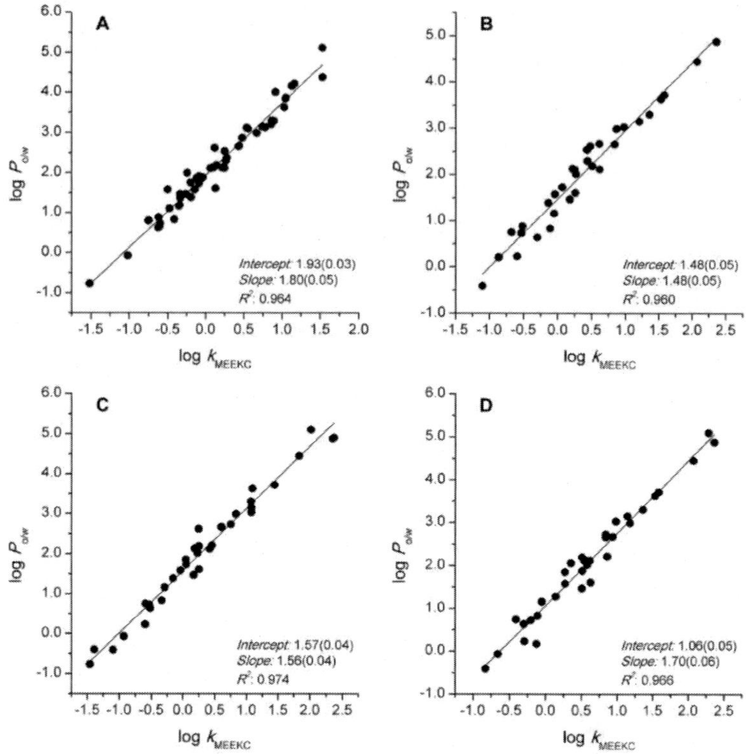

Figure 3. Linear correlations between log $P_{o/w}$ and log k obtained for microemulsions consisting of 8.15% (v/v) 1-butanol, 1.15% (v/v) heptane and (A) 1.3% (w/v) SDS, 10% (v/v) acetonitrile, 20 mM phosphoric acid buffer pH 2.0; (B) 1.3% SDS, 5% acetonitrile, 10 mM phosphate buffer pH 7.4; (C) 1.3% SDS, 10% acetonitrile, 20 mM borate buffer pH 10.0; and (D) 3.3% SDS, 5% acetonitrile, 10 mM phosphate buffer pH 12.0. From ref. [21], with permission.

Going one step forward, Ishihama et al. [18] proposed a microemulsion as running buffer in a capillary electrophoresis format for the estimation of log $P_{o/w}$ from capacity factors. The principle of microemulsion electrokinetic chromatography (MEEKC) is very similar to MEKC, but substituting the micelles by a dispersion of oil droplets in an aqueous buffer. Surfactants and cosurfactants are also needed in order to coat the droplets and reduce the surface tension between oil and water (Figure 2), resulting in the formation of a thermodynamically stable single-phase solution. In the above mentioned work, the microemulsion consisted of 1.44% SDS (surfactant), 6.49% 1-butanol (cosurfactant) and 0.82% heptane (oil) in 0.1 M borate – 0.05 M phosphate, pH 7.0. Since then, microemulsions of different composition (surfactant type and concentration, buffering species and pH ranges from 2 to 12…) have been assayed in order to propose MEEKC as a high throughput and reliable technique for the estimation of log $P_{o/w}$, including different capillaries (fused silica and coated) and mass spectrometry detection [19–21].

If needed, up to 10% acetonitrile can be added to the microemulsion in order to improve the resolution in the upper lipophilic range, with the aim of shifting to some extent the partition into the hydroorganic buffer of very lipophilic compounds, improving the resolution between the analyte peak and that of the droplet marker. Figure 3 shows some examples of linear correlations between log $P_{o/w}$ and log k obtained for compounds with different acid/base properties using microemulsions of different compositions. Notice the broad log $P_{o/w}$ range of application, in most cases between -0.5 and 5.

The procedure for the measurement of lipophilicity from electrokinetic chromatographic methods can be summarized as follows:

1) Calibration of the system: representative compounds of known log $P_{o/w}$ (typically determined by shake-flask or potentiometric methods) are injected as standards in the chromatographic system. A straight line such as log $P_{o/w} = a \cdot \log k + b$ is obtained (Figure 4).
2) Measuring of the mass distribution ratios of solutes with unknown lipophilicity by means of Equation (3). A very lipophilic compound (e.g., dodecanophenone) and a very lipophobic substance (e.g., DMSO) are needed in order to define the migration window.
3) Estimation of log $P_{o/w}$ values from mass distribution ratios by means of the calibration equation stablished in step (1).

Notice that chromatographic methods for lipophilicity determination are mainly intended for unionized solutes, although MEEKC has also been applied to acidic drugs [23]. Charged analytes, besides the solute partition between micelles or oil droplets and the bulk aqueous phase, would additionally migrate inside the capillary under the electric field, and electrostatic interactions might also take place between ionized solutes and charged surfactants in the micelles or oil droplets [24, 25].

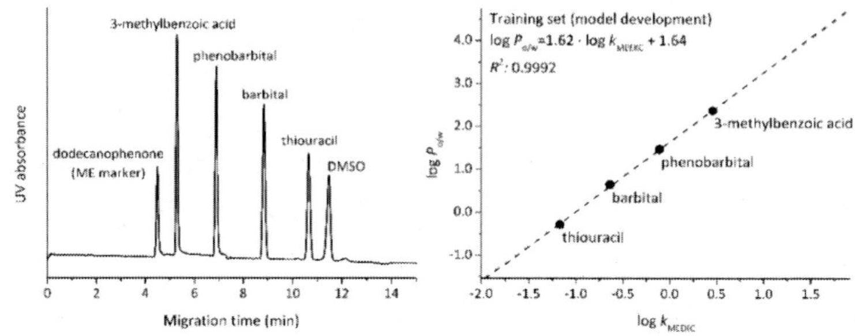

Figure 4. Electropherogram of the calibration mixture with DMSO and dodecanophenone as electroosmotic flow and droplet markers, respectively, and the resulting calibration equation. Microemulsion composition: 1.3% (w/v) SDS, 8.15% (v/v) 1-butanol, 1.15% (v/v) heptane in 20 mM phosphoric acid, pH 2.0. Instrumental conditions: fused silica capillary (50 µm id, 375 µm od, 57.0/48.5 cm of total/effective length), -24 kV (inlet, cathode; outlet, anode), 50 mbar of external pressure (on inlet vial). This procedure was validated for acidic compounds in the log $P_{o/w}$ range -1.54 to 4.75 with a prediction accuracy of ±0.4. From ref. [26] (CC BY 3.0).

MEKC AND MEEKC AS SURROGATE MODELS FOR PHYSICOCHEMICAL AND BIOLOGICAL SYSTEMS

In electrokinetic capillary chromatography the properties of the pseudo-stationary phase can be easily adjusted changing the composition of the running buffer, mainly the type of surfactant. The anisotropic amphiphilic nature of micelles resembles the structure of biomembranes, and allows the exploration of these chromatographic systems as surrogate models for biological processes.

Similarity of Systems: Abraham's Solvation Model

The Abraham's solvation model is a very informative tool based on linear free energy relationships, which provides information about the factors influencing the partition of a solute between two different phases. According to this model, the energy related to a solvation process of a compound (log *SP*) is linearly related to specific interactions between the solute and the solvent molecules of the surrounding phase. These interactions are mainly dispersion ($e{\cdot}E$, interactions through π- and n-electron pairs), dipole-dipole or dipole-induced dipole plus some polarizability interactions ($s{\cdot}S$), solute hydrogen-bond acidity and basicity ($a{\cdot}A$ and $b{\cdot}B$, respectively), and a volume term ($v{\cdot}V$) related to the work of separating solvent molecules to provide a cavity of suitable size for the solute [27].

$$\log SP = c + e{\cdot}E + s{\cdot}S + a{\cdot}A + b{\cdot}B + v{\cdot}V \quad (4)$$

Capital letters in Equation (4) represent the molecular descriptors of a solute, which are constant values greater than or equal to zero, either experimentally measured or calculated from the 2D structure [28]. c is a system constant independent of the solutes, and e, s, a, b, and v are system coefficients, which depend on the solvents (or phases) under consideration. In the case of octanol/water partition (log $P_{o/w}$), these system coefficients measure the differences between both immiscible phases, organic and aqueous. If a chromatographic system is studied, these two condensed phases are the stationary (e.g., liquid C18) and the mobile (e.g., acetonitrile/aqueous buffer) phases. For instance, Abraham [29] used a training set of 613 compounds for the characterization of the octanol/water system, and found the following equation:

$$\log P_{o/w} = 0.088 + 0.562E - 1.054S + 0.034A - 3.460B + 3.814V \quad (R^2 = 0.995) \quad (5)$$

Equation 5 shows that hydrogen-bond basicity and molecular volume of solutes play a key role in the partitioning between octanol and water. Contrarily to octanol, water molecules are excellent hydrogen-bond donors that stabilize solutes with hydrogen-bond acceptor features. This is represented with a negative coefficient of b, which reduces the extent of the partition into the organic phase in favour of the aqueous one, decreasing thus the log $P_{o/w}$ value. The same happens with dipolar and polarizable solutes (negative s), but to a lower extent. At the other end of the scale, just because of intermolecular cohesion of water molecules, a significant amount of energy is required to create a cavity for the solute in water (which is not the case for octanol) and thus larger compounds partition more favourably into the organic phase (positive v). Solutes with π- and n-electron pairs also show a higher degree of interaction with octanol molecules rather than water (positive e). Differences in hydrogen-bond basicity of both solvents are not very relevant (a close to zero).

If the solute property (log SP) is chromatographic retention (log k), positive coefficients mean a higher affinity of the analyte for the stationary phase and thus an increase in retention. On the contrary, negative coefficients show a higher similarity of solutes with the mobile phase and therefore a decrease in chromatographic retention.

With the aim of comparing the similarity of systems characterized by Abraham's solvation model, each one can be considered as a vector in a five-dimensional space (e, s, a, b, and v). Since the measured solute property (log SP) can show different magnitudes (for instance, in Figure 3 the log k range is about -1.5 to 2.5, whereas for log $P_{o/w}$ it is about -1 to 5), it is convenient to normalize the system coefficients dividing by the length of the vector. Once normalized system vectors are obtained, the Euclidean distance (d) between two systems can be calculated as a measure of their similarity. Systems can be considered similar if d is close to zero or at least lower than 0.25 [30].

Comparison between Physicochemical Systems: log $P_{o/w}$ and Chromatography

Table 1 shows the normalized system coefficients for octanol/water partition and different chromatographic approaches, either conventional (HPLC) or electrokinetic (MEEKC and MEKC). Similarity between log $P_{o/w}$ and chromatographic systems is calculated by means of the d parameter. As expected from the previous discussion, the highest similarity was obtained for MEEKC. Normalized coefficients (e_u, s_u, a_u, b_u, and v_u) are very close to those of log $P_{o/w}$, and therefore calculated *d* values are generally below 0.10. Interestingly, variations in the microemulsion pH and the content of acetonitrile, added to improve the resolution of lipophilic analytes, show a minor impact on system coefficients. MEKC with sodium cholate (SC) as surfactant and reversed-phase liquid chromatography show a reasonable similarity with octanol/water, but with significant differences in the hydrogen-bonding acceptor properties (a_u). Whereas this coefficient is slightly positive for log $P_{o/w}$ and MEKC SC, it is clearly negative for liquid chromatography. This observation is consistent with the need of introducing molecular descriptors for hydrogen-bonding acidity of solutes in the accurate measurement of log $P_{o/w}$ from reversed-phase retention factors.

Table 1. Normalized coefficients for some chromatographic systems used for the lipophilicity measurement and their similarity (*d*) to log $P_{o/w}$

Physicochemical system	e_u	s_u	a_u	b_u	v_u	d
log $P_{o/w}$, Octanol/water	0.11	-0.20	0.01	-0.65	0.72	-
MEEKC SDS, pH 2.0	0.09	-0.22	-0.02	-0.63	0.74	0.05
MEEKC SDS, pH 2.0 10%ACN	0.08	-0.19	-0.03	-0.64	0.74	0.05
MEEKC SDS, pH 10.0 10%ACN	0.05	-0.17	0.02	-0.66	0.73	0.06
MEEKC SDS, pH 12.0 5%ACN	0.15	-0.23	0.02	-0.62	0.74	0.06
MEEKC SDS, pH 10.0	0.11	-0.15	-0.03	-0.68	0.71	0.07
MEEKC SDS, pH 7.4 5%ACN	0.12	-0.22	-0.07	-0.61	0.75	0.10
MEKC SC (80 mM)	0.22	-0.22	0.04	-0.62	0.72	0.12
HPLC RP pH 7 40%ACN	0.10	-0.20	-0.11	-0.68	0.69	0.13
HPLC RP pH 7 60%ACN	0.08	-0.23	-0.13	-0.67	0.69	0.15
MEKC SDS (40 mM) pH 7.0	0.17	-0.18	-0.08	-0.51	0.82	0.21
MEKC SDC (40 mM) pH 8.0	0.28	-0.27	0.02	-0.55	0.74	0.22
MEKC LDS (40 mM) pH 7.0	0.18	-0.19	-0.10	-0.49	0.82	0.23
MEKC TTAB (20 mM) pH 7.0	0.24	-0.16	0.20	-0.63	0.69	0.24
MEKC HTAB (20 mM) pH 7.0	0.28	-0.19	0.21	-0.61	0.68	0.27
MEKC LPFOS (40 mM) pH 7.0	-0.05	-0.11	-0.40	-0.21	0.89	0.65

MEEKC systems contain 1.3% SDS, 8.15% 1-butanol and 1.15% heptane [21].
MEKC data calculated from refs. [31,32]. Abbreviations for surfactants at the end of the chapter.
Reversed-phase HPLC column: XTerra RP18 (Waters, Milford, MA) [30].

The nature of the surfactant plays a fundamental role in defining the properties of MEKC systems. When compared with octanol/water, the lowest *d* value was obtained for micelles prepared from the bile salt SC (0.12), followed by SDS (0.21), and lithium perfluorooctane sulfonate (LPFOS) was responsible for the highest dissimilitude (0.65).

Electrokinetic Chromatography as Surrogate Models for Biological Systems

Since micelles and microemulsions reasonably mimic the properties of cell membranes, MEKC and MEEKC are potential models for biological systems. This is especially relevant when biological activity has to be determined *in vivo*, which might be time consuming, expensive and a non-neglible source of ethical concerns regarding animal experimentation. It is easy to modify the behavior of the pseudo-stationary phase in electrokinetic chromatography just by changing the surfactant employed in the micelles or the oil droplets, or even preparing mixtures of them. Then they can be characterized by means of the Abraham's solvation model and compared with a biological system of interest with the aim of finding the best possible match. The solute property representative of a particular biopartitioning system depends on its nature and the measurement scale traditionally used. For instance, in the case of skin partition it can be log K_{sc}, whereas for non-specific water toxicity towards fish or water flea species -log LC_{50} can be employed, or -log MIC for bacterial growth inhibition. Figure 5 is an example of correlations between biologically relevant properties and capacity factors (log *k*) obtained in a MEKC system containing STC as surfactant. A linear relationship is always desirable, similar to that obtained for log *k* in MEEKC and log $P_{o/w}$.

Hidalgo [33] compiled the system constants of 28 biopartitioning systems characterized by means of the solvation parameter model (Eq. (4)), and compared them with 92 physicochemical partitioning systems, including immiscible organic solvent/water, conventional liquid and electrokinetic chromatographies. Table 2 shows the goodness of some

MEKC/MEEKC systems as surrogate models of biological activity, expressed as *d* values. Minimum distance for each reported biopartitioning system is marked in bold. Thus, MEKC AOT seems to be the best option for skin partition studies and MEKC STDC for tadpole narcosis. The best surrogate for aquatic toxicity towards to fathead minnow and guppy fishes is MEKC SDS-Brij 35, but the best option for fish medaka high-eyes is MEKC TTAB.

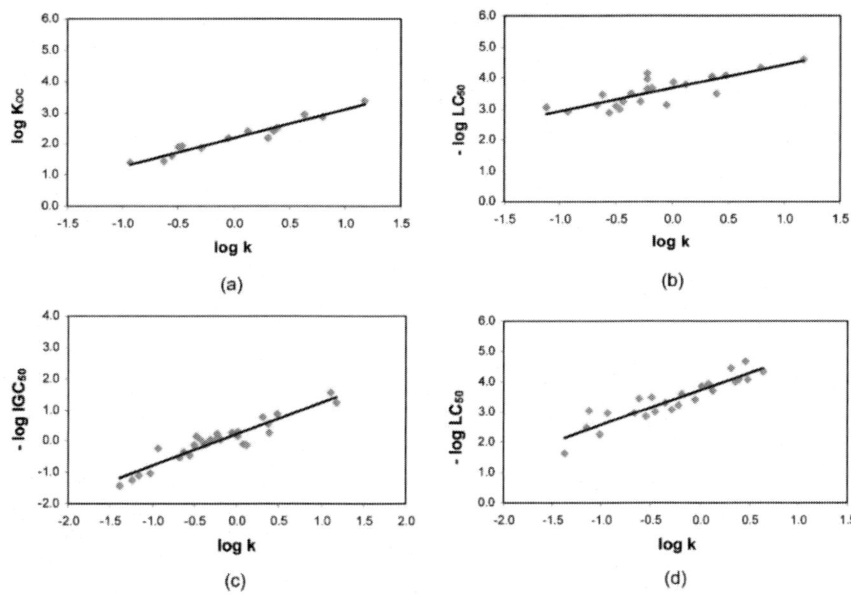

Figure 5. Plots of experimental (a) soil−water partition logarithm (log K_{OC}), (b) median mortality lethal concentration logarithm (-log LC_{50}) for the guppy fish, (c) median inhibitory growth cell concentration logarithm (-log IGC_{50}) for the *Tetrahymena Pyriformis* protozoa, and (d) median mortality lethal concentration logarithm (-log LC50) for the fathead minnow fish vs experimental retention factor logarithm (log *k*) in MEKC STC (50 mM STC in 10 mM NaH_2PO_4−10 mM $Na_2B_4O_7$ aqueous buffer, pH 8). From ref. [34], with permission.

Growth inhibition towards bacterium *Porphyromonas gingivalis* can be studied with MEKC HTAB, but *Selenomonas artemidis* is more similar to MEKC DHAB. Interestingly, some MEKC systems prepared with different surfactants show similar distances to the same biopartitioning systems, such as ALE and SLN, SDCV and SLSA or STC and STDC, but some surfactants like LPFOS lead to chromatographic systems very different to any of the reported biological ones. Just for the sake of comparison, *d* distances with those systems reported to be very similar to octanol-water partition (MEEKC SDS and MEKC SC) are also presented in Table 2.

Table 2. Similarity (*d*) between some biological and electrokinetic chromatographic partitioning systems [33]

				Non-specific aquatic toxicity towards:					**Growth inhibition:**	
	Surfactant	Skin partition [35]	Tadpole narcosis [36]	Fish fathead minnow [37]	Fish guppy [37]	Fish medaka high-eyes [37]	Water flea *Daphnia magna* [38]	Protozoa *Tetrahymena pyriformis* [38]	Bacterium *Porphyromonas gingivalis* [39]	Bacterium *Selenomonas artemidis* [39]
MEKC	ALE [40]	0.22	0.12	0.15	0.11	0.20	0.17	0.11	0.26	0.26
MEKC	AOT [41]	0.10	0.17	0.11	0.13	0.31	0.11	0.12	0.36	0.33
MEKC	DHAB [42]	0.26	0.24	0.21	0.18	0.14	0.24	0.22	0.20	0.07
MEKC	HTAB [31, 32]	0.33	0.20	0.30	0.23	0.10	0.32	0.29	0.08	0.14
MEKC	LMT [40]	0.21	0.10	0.15	0.09	0.20	0.16	0.10	0.25	0.26
MEKC	LPFOS [31, 32]	0.74	0.65	0.75	0.72	0.81	0.72	0.71	0.82	0.91
MEKC	SC [43]	0.29	0.15	0.31	0.23	0.23	0.31	0.30	0.20	0.28
MEKC	SDCV [44]	0.21	0.08	0.19	0.15	0.27	0.19	0.17	0.30	0.32
MEEKC	SDS [45]	0.16	0.15	0.18	0.18	0.33	0.18	0.19	0.36	0.35
MEKC	SDS-Brij 35 [46]	0.13	0.13	0.09	0.07	0.23	0.12	0.10	0.28	0.24
MEKC	SLN [40]	0.23	0.10	0.16	0.12	0.20	0.18	0.12	0.25	0.27
MEKC	SLSA [44]	0.19	0.10	0.16	0.12	0.27	0.15	0.12	0.31	0.33
MEKC	STC [47]	0.18	0.09	0.18	0.10	0.24	0.17	0.16	0.28	0.31
MEKC	STDC [48]	0.20	0.07	0.20	0.13	0.24	0.19	0.18	0.27	0.31
MEKC	TTAB [31, 32]	0.29	0.18	0.25	0.19	0.09	0.27	0.24	0.12	0.12

Clearly, there is not a universal chromatographic system able to mimick any biological system, but the versatility of electrokinetic cromatographic approaches makes them excellent candidates in the search of surrogate models of biological activity.

Conclusion

Partitioning systems, being either physicochemical or biological, can be potentially modeled by electrokinetic chromatographic techniques in capillary format. Lipophilicity can be measured by MEEKC SDS and expressed in the widely used log $P_{o/w}$ scale, with the advantages of reduced analysis times (from days/hours to minutes) and lower sample purity requirements. The nature of the surfactant in MEKC plays a key role on the properties of micelles, making it possible to find a micelle composition able to mimick to some extent biopartitioning systems. For instance, it is the case of tadpole narcosis and MEKC STDC, non-specific aquatic toxicity towards fathead minnow and guppy fishes and MEKC SDS-Brij 35, or growth inhibition towards bacterium *Selenomonas artemidis* and MEKC DHAB.

References

[1] Arnott, J. A., S. L. Planey. (2012). The influence of lipophilicity in drug discovery and design, *Expert Opin. Drug Discov.* 7:863–875. doi:10.1517/17460441.2012.714363.

[2] Di, L., E. H. Kerns, Chapter 5 - *Lipophilicity. Drug-Like Properties: concepts, structure design and methods from ADME to toxicity optimization.*, Academic Press, 2016. doi:10.1016/B978-0-12-801076-1.00005-8.

[3] Avdeef, A. *Absorption and drug development*, Second, John Wiley & Sons, Inc., Hoboken, NJ, US, 2012. doi:10.1002/9781118286067.

[4] OECD, *Test No. 107: Partition Coefficient (n-octanol/water): Shake Flask Method*, OECD Publishing, 1995. doi:10.1787/9789264069626-en.

[5] EPA, *Product Properties Test Guidelines. OPPTS 830.7550.* Partition coefficient (*n*-octanol/water), shake-flask method, U.S. Environmental Protection Agency, Prevention, Pesticides and Toxic Substances, United States, 1996. http://fedbbs.access.gpo.gov/library/epa_830/830-7550.pdf.

[6] Andrés, A., M. Rosés, C. Ràfols, E. Bosch, S. Espinosa, V. Segarra, J. M. Huerta. (2015). Setup and validation of shake-flask procedures for the determination of partition coefficients (log *D*) from low drug amounts, *Eur. J. Pharm. Sci.* 76:181–191. doi:10.1016/j.ejps.2015.05.008.

[7] ECHA, *Guidance on Information Requirements and Chemical Safety Assessment*. Chapter R.7a: Endpoint specific guidance, 2016. doi:10.2823/2611.

[8] Avdeef, A. (1992). pH-Metric log P. Part I. Difference Plots for Determining Ion-Pair Octanol-Water Partition Coefficients of Multiprotic Substances, *Quant. Struct. Relationships.* 11:510–517. doi:10.1002/qsar.2660110408.

[9] Avdeef, A. (1993). pH-Metric log P. II: Refinement of Partition Coefficients and Ionization Constants of Multiprotic Substances, *J. Pharm. Sci.* 82:183–190. doi:10.1002/jps.2600820214.

[10] Port, A., M. Bordas, R. Enrech, R. Pascual, M. Rosés, C. Ràfols, X. Subirats, E. Bosch. (2018). Critical comparison of shake-flask, potentiometric and chromatographic methods for lipophilicity evaluation (log $P_{o/w}$) of neutral, acidic, basic, amphoteric, and zwitterionic drugs, *Eur. J. Pharm. Sci.* 122:331–340. doi:10.1016/j.ejps.2018.07.010.

[11] Leo, A., C. Hansch, D. Elkins. (1971). Partition coefficients and their uses, *Chem. Rev.* 71:525–616. doi:10.1021/cr60274a001.

[12] Donovan, S. F., M. C. Pescatore. (2002). Method for measuring the logarithm of the octanol-water partition coefficient by using short octadecyl-poly(vinyl alcohol) high-performance liquid

chromatography columns, *J. Chromatogr. A.* 952:47–61. doi:10.1016/S0021-9673(02)00064-X.

[13] Valko, K., C. Du, C. Bevan, D. Reynolds, M. Abraham. (2001). Rapid method for the estimation of octanol/water partition coefficient from gradient RP-HPLC retention and a hydrogen bond acidity term, *Curr. Med. Chem.* 8:1137–1146. doi:10.2174/0929867013372643.

[14] Pallicer, J. M., R. Pascual, A. Port, M. Rosés, C. Ràfols, E. Bosch (2013). The contribution of the hydrogen bond acidity on the lipophilicity of drugs estimated from chromatographic measurements, *Eur. J. Pharm. Sci.* 48:484–493. doi:10.1016/j.ejps.2012.12.008.

[15] Terabe, S., K. Otsuka, K. Ichikawa, A. Tsuchiya, T. Ando. (1984). Electrokinetic Separations with Micellar Solutios and Open-Tublar Capillaries, *Anal. Chem.* 56: 111–113.

[16] Yang, S., J. G. Bumgarner, L. F. R. Kruk, M. G. Khaledi. (1996). Quantitative structure-activity relationships studies with micellar electrokinetic chromatography: Influence of surfactant type and mixed micelles on estimation of hydrophobicity and bioavailability, *J. Chromatogr. A.* 721:323–335. doi:10.1016/0021-9673(95) 00773-3.

[17] Holzgrabe, U., S. Schmitt, F. Wienen, Micellar Electrokinetic Chromatography of Aminoglycosides, in: P. Schmitt-Kopplin, Capill. *Electrophor. Methods Protoc.*, Springer New York, New York, NY, 2016: 77–90. doi:10.1007/978-1-4939-6403-1_5.

[18] Ishihama, Y., Y. Oda, K. Uchikawa, N. Asakawa. (1995). Evaluation of solute hydrophobicity by microemulsion electrokinetic chromatography, *Anal. Chem.* 67:1588–1595. doi:10.1002/ elps.200410141.

[19] Poole, S. K., D. Durham, C. Kibbey. (2000). Rapid method for estimating the octanol-water partition coefficient (log P_{ow}) by microemulsion electrokinetic chromatography, *J. Chromatogr. B Biomed. Sci. Appl.* 745:117–126. doi:10.1016/S0378-4347(00)00072-4.

[20] Henchoz, Y., S. Romand, J. Schappler, S. Rudaz, J. L. Veuthey, P. A. Carrupt. (2010). High-throughput log *P* determination by MEEKC

coupled with UV and MS detections, *Electrophoresis*. 31:952–964. doi:10.1002/elps.200900540.

[21] Subirats, X., H. P. Yuan, V. Chaves, N. Marzal, M. Rosés. (2016). Microemulsion electrokinetic chromatography as a suitable tool for lipophilicity determination of acidic, neutral, and basic compounds, *Electrophoresis.* 37:2010–2016. doi:10.1002/ELPS.2016 00080.

[22] Yang, L., X. Xie, J. Zhang, G. Sun. (2016). Microemulsion electrokinetic chromatography in combination with chemometric methods to evaluate the holistic quality consistency and predict the antioxidant activity of Ixeris sonchifolia (Bunge) Hance Injection, *PLoS One*. 11:1–19. doi:10.1371/journal.pone.0157601.

[23] Fernández-Pumarega, A., S. Amézqueta, E. Fuguet, M. Rosés. (2018). Feasibility of the estimation of octanol-water distribution coefficients of acidic drugs by microemulsion electrokinetic chromatography, *ADMET DMPK* 6: 55–60. doi:10.5599/admet.6.1.510.

[24] Khaledi, M. G., S. C. Smith, J. K. Strasters. (1991). Micellar electrokinetic capillary chromatography of acidic solutes: migration behavior and optimization strategies., *Anal. Chem.* 63:1820–1830. doi:10. 1021/ac00017a029.

[25] Téllez, A., E. Fuguet, M. Rosés. (2007). Comparison of migration models for acidic solutes in micellar electrokinetic chromatography, *J. Chromatogr. A*. 1139:143–151. doi:10.1016/j.chroma.2006. 11.004.

[26] Subirats, X., L. Redón, M. Rosés. (2018). Lipophilicity determination of acidic compounds: MEEKC as a reliable high-throughput methodology, *ADMET DMPK*. 6:153–161. doi:10.5599/admet .491.

[27] Abraham, M. H. (1993). Scales of solute hydrogen-bonding: their construction and application to physicochemical and biochemical processes, *Chem. Soc. Rev*. 22:73. doi:10.1039/cs9932200073.

[28] Ulrich N. E., S. Brown, T. N., Watanabe, N., Bronner, G., Abraham, M. H., Goss, K. U. (2017), *UFZ-LSER database v 3.2* [Internet]. http://www.ufz.de/lserd.

[29] Abraham, M. H., H. S. Chadha, G. S. Whiting, R. C. Mitchell. (1994). Hydrogen Bonding. 32. An Analysis of Water-Octanol and Water-

Alkane Partitioning and the Δlog P Parameter of Seiler, *J. Pharm. Sci.* 83:1085–1100. doi:10.1002/jps.2600830806.

[30] Lázaro, E., C. Ràfols, M. H. Abraham, M. Rosés. (2006). Chromatographic estimation of drug disposition properties by means of immobilized artificial membranes (IAM) and C18 columns, *J. Med. Chem.* 49: 4861–4870. doi:10.1021/jm0602108.

[31] Fuguet, E., C. Ràfols, E. Bosch, M. H. Abraham, M. Rosés. (2002). Solute-solvent interactions in micellar electrokinetic chromatography: III. Characterization of the selectivity of micellar electrokinetic chromatography systems, *J. Chromatogr. A.* 942:237–248. doi:10.1016/S0021-9673(01)01383-8.

[32] Fuguet, E., C. Ràfols, E. Bosch, M. H. Abraham, M. Rosés. (2009). Erratum to "Solute–solvent interactions in micellar electrokinetic chromatography. III. Characterization of the selectivity of micellar electrokinetic chromatography systems" [*J. Chromatogr. A* 942 (2002) 237–248], *J. Chromatogr. A.* 1216:6877–6879. doi:10.1016/j.chroma.2009.07.063.

[33] Hidalgo-Rodríguez, M. *Estimació de propietats de biopartició mitjançant mesures cromatogràfiques* [*Estimation of biopartitioning properties by means of chromatographic measurements*], University of Barcelona, 2013. http://hdl.handle.net/10803/128267.

[34] Hidalgo-Rodríguez, M., E. Fuguet, C. Ràfols, M. Rosés. (2010). Estimation of Biological Properties by Means of Chromatographic Systems : Evaluation of the Factors That Contribute to the Variance of Biological - Chromatographic Correlations, *Anal. Chem.* 82:10236–10245.

[35] Abraham, M. H., F. Martins. (2004). Human Skin Permeation and Partition: General Linear Free Energy Relationship Analyses, *J. Pharm. Sci.* 93:1508–1523. doi:10.1002/jps.20070.

[36] Abraham, M. H., C. Ràfols. (1995). Factors that influence tadpole narcosis. An LFER analysis, *J. Chem. Soc. Perkin Trans.* 2:1843–1851. doi:10.1039/P29950001843.

[37] Hoover, K. R., W. E. Acree, M. H. Abraham. (2005). Chemical toxicity correlations for several fish species based on the Abraham

solvation parameter model, *Chem. Res. Toxicol.* 18:1497–1505. doi:10. 1021/tx050164z.

[38] Hoover, K. R., K. B. Flanagan, J. Acree William E, M. H. Abraham. (2007). Chemical toxicity correlations for several protozoas, bacteria, and water fleas based on the Abraham solvation parameter model, *J. Environ. Eng. Sci.* 6:165–174. doi:10.1139/s06-041.

[39] Mintz, C., W. E. Acree, M. H. Abraham. (2006). Correlation of Minimum Inhibitory Concentrations Toward Oral Bacterial Growth Based on the Abraham Model, *QSAR Comb. Sci.* 25:912–920. doi:10.1002/qsar.200630011.

[40] Trone, M. D., M. G. Khaledi. (2000) Influence of ester and amide-containing surfactant headgroups on selectivity in micellar electrokinetic chromatography, *Electrophoresis.* 21:2390–2396. doi:10.1002/1522-2683(20000701)21:12<2390::AID-ELPS239 0>3.0 .CO;2-K.

[41] Klotz, W., M.. Schure, J.. Foley. (2002). Rapid estimation of octanol–water partition coefficients using synthesized vesicles in electrokinetic chromatography, *J. Chromatogr. A.* 962:207–219. doi:10.1016/ S0021-9673(02)00352-7.

[42] Agbodjan, A. A., M. G. Khaledi. (2003). Study of solute partitioning into cationic vesicles of dihexadecyldimethylammonium bromide using electrokinetic chromatography, *J. Chromatogr. A.* 1004:145–153. doi:10.1016/S0021-9673(03)00853-7.

[43] Hidalgo-Rodríguez, M., E. Fuguet, C. Ràfols, M. Rosés. (2010). Solute–solvent interactions in micellar electrokinetic chromatography: VII. Characterization of sodium cholate–sodium deoxycholate mixed-micellar systems, *J. Chromatogr. A.* 1217:1701–1708. doi:10.1016/ j.chroma.2010.01.001.

[44] Trone, M. D., M. G. Khaledi. (1999). Characterization of Chemical Selectivity in Micellar Electrokinetic Chromatography. 4. Effect of Surfactant Headgroup, *Anal. Chem.* 71:1270–1277. doi:10. 1021/ac9809736.

[45] Abraham, M. H., C. Treiner, M. Rosés, C. Ràfols, Y. Ishihama. (1996). Linear free energy relationship analysis of microemulsion

electrokinetic chromatographic determination of lipophilicity, *J. Chromatogr. A*. 752: 243–249. doi:10.1016/S0021-9673(96) 00518-3.

[46] Rosés, M., C. Ràfols, E. Bosch, A. M. Martínez, M. H. Abraham. (1999). Solute-solvent interactions in micellar electrokinetic chromatography. Characterization of sodium dodecyl sulfate-Brij 35 micellar systems for quantitative structure-activity relationship modelling, *J. Chromatogr. A*. 845:217–226. doi:10.1016/ S0021-9673(99)00147-8.

[47] Hidalgo-Rodríguez, M., E. Fuguet, C. Ràfols, M. Rosés. (2012). Modeling nonspecific toxicity of organic compounds to the fathead minnow fish by means of chromatographic systems, *Anal. Chem.* 84:3446–3452. doi:10.1021/ac2034453.

[48] Poole, S. K., C. F. Poole. (1997). Characterization of Surfactant Selectivity in Micellar Electrokinetic Chromatography, *Analyst.* 122:267–274. doi:10.1039/a605799c.

In: Capillary Electrophoresis … ISBN: 978-1-53615-223-4
Editors: O. Núñez and X. Subirats

Chapter 7

CAPILLARY ELECTROPHORESIS IN THE PHARMACEUTICAL AND BIOTECH INDUSTRY

Ewoud van Tricht[1], Lars Geurink[1,2] and Cari E. Sänger – van de Griend[2,3,*]

[1]Analytical Development, Janssen Vaccines and Prevention BV, Leiden, the Netherlands
[2]Department of Medicinal Chemistry, Uppsala University, Uppsala, Sweden
[3]Kantisto BV, Baarn, the Netherlands

ABSTRACT

Capillary Electrophoresis (CE) has led a turbulent life in the pharmaceutical industry. Application of the CE techniques have continuously been increasing, but after the initial enthusiastic phase and fast adaptation, there was a decline in use at the start of the 21st century for small molecules. In this chapter the implementation and use of the CE

---

[*] Corresponding Author's Email: cari.sanger@kantisto.nl.

applications in the pharmaceutical and biotechnology industry in the early twenty-first century are discussed. The requirements for CE to become a well-established technique for the analysis of therapeutic proteins are discussed as they also set the requirements for further development. Relevant old and new applications in industry illustrate the versatility of use of the CE techniques in industry. The application of Analytical Quality by Design (AQbD) on capillary electrophoresis method development, validation, implementation, and life cycle management is reviewed. The chapter ends with expectations of and hopes for the future use of the CE techniques. If the CE community keeps sharing good practice and good applications, CE will have a bright future in the pharmaceutical and biotech industry.

ABBREVIATIONS

ACE	Affinity CE
AQbD	Analytical Quality by Design
ATP	Analytical target profile
ATPS	9-Aminopyrene-1,4,6-trisulfonic acid
BGE	Background electrolyte solution
BLA	Biologics License Application
CE	Capillary electrophoresis
CE-SDS	Capillary electrophoresis-sodium dodecyl sulfate
CEC	Capillary electrochromatography
CDT	Carbohydrate deficient transferrin
CFR	Code of Federal Regulations
CGE	Capillary gel electrophoresis
CHES	N-Cyclohexyl-2-aminoethanesulfonic acid
cIEF	Capillary isoelectric focusing
CZE	Capillary zone electrophoresis
C^4D	Capacity coupled contactless conductivity detection/ detector
DOE	Design of Experiments
DMPK	Drug metabolism and pharmacokinetics
EOF	Electroosmotic flow
EPO	Erythropoietin

FIDA	Flow-induced dispersion analysis
FISH	Fluorescence in situ hybridization
FDA	US Food and Drug Administration
GMP	Good Manufacturing Practice
HPLC	High-performance liquid chromatography
HPMC	Hydroxypropylmethylcellulose
HRV	Human Rhinovirus
IACE	Immunoaffinity CE
icIEF	Imaging cIEF
LC	Liquid chromatography
LIF	Laser-induced fluorescence detection
LOQ	Quantitation limit
mAb	Monoclonal antibody
MEEKC	Micro-emulsion electrokinetic chromatography
MEKC	Micellar electrokinetic chromatography
MS	Mass spectrometry
NCE	New chemical entity
QbD	Quality by Design
QC	Quality Control
PhEur	European pharmacopoeia
pI	Isoelectric point
PAT	Process analytical technology
PVA	Polyvinyl alcohol
RSD	Relative standard deviation
SDS	Sodium dodecyl sulfate
SDS-PAGE	SDS polyacrylamide gel electrophoresis
SOP	Standard Operating Procedure
TCEP	Tris(2-carboxyethyl) phosphine
TETA	Triethylenetetramine
tITP	Transient isotachophoresis
TMV	Tobacco mosaic virus
TOF	Time-of-flight
TRIS	Tris(hydroxymethyl)aminomethane
USP	US Pharmacopoeia

UV	Ultraviolet
VLP	Virus like particles
μCE-SDS	Chip-sized CE-SDS

INTRODUCTION

Professor Arne Tiselius, winner of the Nobel Prize in Chemistry in 1948 for his fundamental work on electrophoresis said "… the need for specific and gentle methods of separation in the high molecular field is great, especially with respect to the enormous multitude of substances occurring in nature which are similar chemically but dissimilar with respect to function and activity" [1]. A crucial step to this aim was made by Tiselius' student, professor Stellan Hjertén, who set the foundations for capillary electrophoresis [2] that were later translated into practice by professor Jim Jorgensson [3]. Already in its early days, CE was used for the analysis of (bio)pharmaceuticals. Several of the methods that were developed in those early days are still in use for the quality control of small-molecule and biotherapeutic drugs. Since the 1990s, CE methods have also been introduced in the pharmacopoeias. Overviews can be found in [4-6]. The current increase in use is mainly attributed to biotherapeutics. Continuous application development and renewed instrumental development are key, as discussed below. This is reflected in the inclusion of CE methodologies in the specifications for biotherapeutics, from 35% of the products approved before 2009 to > 90% of the products approved in 2016 – 2017 by the US Food and Drug Administration (FDA) [7].

HOW DID CE BECOME A WELL-ESTABLISHED TECHNIQUE?

Let us look at what it takes to implement a new technique into the pharmaceutical industry by taking CE for monoclonal antibodies as an illustrative example. Four requirements are discussed below. This does not mean that this is the order in which the requirements need to be fulfilled. Rather, control over all aspects needs to be gained simultaneously, and one aspect cannot improve without the other.

Requirement 1: Instrumentation

Implementation of a new technology in a highly regulated industry requires the availability of sound equipment. In the late 1980s and early 1990s, the first commercial instruments were introduced. Some of these were hardly better than home-build equipment, others were equipped with autosamplers, appropriate detection units and a pc-based software interface. Needless to say, that the latter thrived better within industry. At the turn of the century, there were only a few instruments that were suitable for a good manufacturing practice (GMP) environment. This was partly based on the Code of Federal Regulations (CFR) part 11 (in)compliance of the software and partly to instrumental characteristics such as injection precision, detection sensitivity and temperature control. Temperature control turned out to be fundamental for good precision. If the temperature fluctuates over a sequence of injections, the viscosity of the electrolyte solution will fluctuate and thus the migration times. Precise temperature control requires high-speed forced-air cooling or liquid cooling. At the turn of the century when there were only a few suitable GMP instruments on the market, instrument development was almost standing still, perhaps due to the lack of competition. Lately, however, there is an increase in instrument development activities and with the firm implementation of the CE technique in the biopharmaceutical market, companies are swift to respond and test new instrumentation. A quick survey in industry showed that most companies have multiple types of instruments available (Figure 1). These include some of the older working horses that are still going strong, such as the Agilent 1600 HP^{3D}CE, the Bioanalyzer and the BeckmanCoulter MDQ. A number of the early instruments showed a relative high rate of invalid analyses, especially for capillary electrophoresis-sodium dodecyl sulfate

(CE-SDS) applications [8, 9]. Improvement of the instrument design, next to more intensive training, has resolved most of these issues, which is paramount for implementation in a quality control (QC) lab. Another trend we see is that newly developed equipment often is application dedicated, such as isoelectric focusing (iCE, Maurice, CEInfinite), CE-SDS (Maurice and chip-based instruments) and capillary western blotting (Wes, Sally Sue, Peggy Sue). Generally, the disadvantage of application instrumentation is that the applications are more or less fixed, and there is little to optimize for the molecule at hand, even if the fixed settings are suboptimal. The advantage of application-dedicated equipment is easiness of implementation. They are usually easier to maintain and handle, which is especially useful in a QC environment. Notably, development of dedicated instrumentation is a sign that implementation in industry has reached such a level, that the development of dedicated equipment is worth the effort.

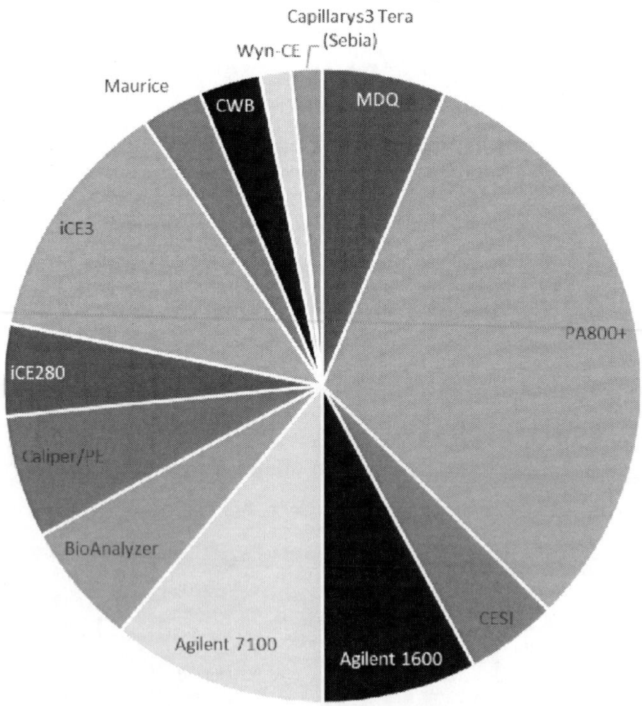

Figure 1. Survey results: CE Equipment in use in industry.

Requirement 2: Relevant Applications and Compendial Methods

Another requirement of implementation of a new technology is the availability of several to-the-point applications. For protein therapeutics, the development of CE-SDS, the capillary format of sodium dodecyl sulfate polyacrylamide gel electrophoresis (SDS-PAGE), was crucial. The application of SDS-PAGE was well developed and a major technique for the characterization of proteins. The introduction of capillary electrophoresis meant that quantitative results could be achieved, without the need of staining and destaining of the proteins, with better precision and higher throughput (see also [10]). A review from Guttman [11] describes the transition well. This paper also makes a comparison between SDS-PAGE and CE-SDS for 65 proteins. Very similar separations, molecular weight determinations and precisions were found. The CE-SDS method is currently a platform method for the characterization of mAbs and serves very well as a general purity test [12]. The platform methodology is so firmly anchored in industry, that also the second requirement for implementation is fulfilled, that is, acceptance of CE methods as compendial methods. The CE-SDS methodology is taken up in the pharmacopoeias, such as in the US Pharmacopeia (USP) general chapter <129> on monoclonal antibodies [13].

The other gel application for proteins that was translated to the capillary format is isoelectric focusing. Capillary isoelectric focusing (cIEF) and imaged cIEF (icIEF) are now well-established within the biopharmaceutical industry and many companies use both technologies next to each other. Intercompany studies showed that cIEF and icIEF are suitable for the charge heterogeneity determination both in process development as well as in quality control [14, 15].

The third application that illustrates how CE is being implemented, is the use of capillary zone electrophoresis (CZE) for protein charge heterogeneity. The use of CZE for charge heterogeneity was given real momentum with the publication of Pfizer of a method based on high concentrations of 6-aminocaproic acid at pH 5.7 [16]. This method now serves as a platform method for mAbs with pIs ranging from 7.5 – 9.5. The method was published and presented at major CE symposia at a point in time when multiple companies were familiar with CE instrumentation and kit-based methodologies, and were now open for further exploration of the potentials of the technique. Indeed, the method quickly gained popularity and already within a few years an intercompany collaboration study was feasible [17]. Examples like this clear the way for further exploitation of CZE for protein characterization, as in our opinion we have not reached the full potential of CE for proteins yet.

Requirement 3: Identification of Good Working Practices

Parallel to the development of key applications, a technique's good working practice needs to be identified. This work was done by the whole CE community, with a special thanks to the industrial meetings such as CEPharm [18] and local user group meetings. Pivotal was also the work of industrial scientists presenting workshops and publishing in assessible book chapters, columns and user notes (e.g., [19-30]. Please note that this list can never be complete, and that the majority of the working practices described here were discussed at user meetings before being published. Nevertheless, a special contribution is due to Kevin Altria. He was not only an early adopter of CE at GlaxoSmithKline, he also channeled the CE good working practices into numerous columns such as the CE Currents series in LCGC [22].

Requirement 4: Training

Al above mentioned requirements for implementation come together in training of the scientists and technicians in the lab. If they are not being taught the governing principles and good working practices of the techniques they use, all above efforts are to no avail. Even if we try to make our methodologies robust, easy to use and fools-prove, training is still important so that the scientist can decide whether the platform method is appropriate, needs adjustment or further development or is not applicable at all. Many issues that arise with day-to-day application of the techniques stems from lack of appropriate training and the use of platform methodologies beyond the scopes applicable. This then creates a false feeling of a technique being difficult or problematic. If however training includes both fundamentals and good working practices and a technique is applied appropriately, it can blossom to its full potential and stimulates creativity to further develop and use a technique at full.

ILLUSTRATIVE APPLICATIONS

A quick survey under CE-users within industry shows that almost all companies work on protein pharmaceuticals, varying from mAbs and hormones to all kind of protein modifications such as pegylated proteins, bispecific antibodies, nanobodies, enzymes, fusion proteins etc. The major applications are CZE and CE-SDS, closely followed by cIEF and icIEF, predominantly for analysis of purity and identity. This fits well with the data presented by Marjorie Shapiro [30], who concluded that although CE-based methods for release and/or stability were used since the 1990s, they became routine in the past 5 – 10 years. CE methods were included in > 95% of the submissions of biologics license applications (BLAs) approved during 2016 – 2017.

Although the CZE publications from industry are stable over the years, there is a shift in the application area. Both the decrease in CE publications on chemicals and drugs and the increase in CE publications in biologicals follow the trend of the emerging biopharmaceutical sector in contrast to the traditional pharmaceutical sector [31]. Recent applications are more for biopharmaceuticals than for small-molecule drugs. Zhu and Scriba conclude in a recent review that for small-molecule drugs, CE is more often used in the academic lab than in the pharmaceutical industry. Reasons mentioned include lack of training and more method variables in CE [32]. However, one should be careful to draw conclusions from the number of papers as many methods developed are never published [5]. In our short survey, all kinds of analytes were mentioned, although proteins in all variations were most prevalent (Table 1). In effect, the applications mentioned cover the whole range from small ions to very large molecules and aggregates such as nucleotides and viruses. The companies participating in the survey have any number from none to over 100 Standard Operating Procedures (SOPs) for these applications. Furthermore, CE is not only applied for pharmaceutical analysis in industry, also bioanalysis and biomarker discovery were mentioned.

Table 1. Different analytes and proteins in all variations

Proteins (Antibodies, protein modifications, pegylated proteins, hormones, hemoglobin, CDT, enzymes)
Peptides, glycopeptides
Ions (counter ions, cations, anions)
Small-molecule drugs, small organic molecules
Carbohydrates, glycans, glycoconjugates
Polysaccharides, glycopolysaccharides
Virus, vaccines
DNA, RNA
Metabolites
Amines, amides
Counterfeit detection

We would like to mention a few illustrative examples of the use of CE in industry here. The purpose of these examples is to illustrate the wide variety of CE techniques, types of analysis and types of analytes used within the (bio)pharmaceutical industry. The intention is not to be a complete overview, rather to inspire the reader.

Small-Molecule Pharmaceuticals

Assay and Purity Determinations

Several different modes of CE have been used for assays and for purity determinations. "Pure" capillary zone electrophoresis has, in all its simplicity, strong separation capabilities and CZE applications consequently cover a large portion of the industrial publications on CE. In CZE, separation is based on charge/size ratios. In general, the publication rate on CZE is stable throughout the years, although one could observe a little peak in the period 1996 – 2002. CZE has been used for drugs, inorganic ions, nucleotides, and recently also increasingly for proteins, peptides and other biological therapeutics. Glaxo was one of the front runners of the use of CE for small-molecule pharmaceuticals, and as early as 1989 on the 1st International Symposium on High-Performance Capillary Electrophoresis, Kevin Altria demonstrated the use of a simple CZE method for the separation of an antidepressant and its impurity [33].

In an Organon paper, validation of a QC method for release and stability of an antidepressant was compared with the existing liquid chromatography (LC) method. The final CZE method proved fit for purpose and robust [34]. The paper also describes what we now recognize as normal development steps in good CE practice, such as the screening of sample extraction solvent and its effect on the current trace as well as the importance of using an internal standard for good injection precision.

Recently, capacity coupled contactless conductivity detection (C^4D) is increasingly being used for detection in CE. C^4D can be more sensitive than indirect UV detection but still cannot compete with ultraviolet (UV) or mass spectrometry (MS) detection for analytes with chromophores [32]. However, a C^4D detector is a simple and cheap solution in situations where money is constrained, such as in underdeveloped and developing countries. Also in those countries, quality control of drugs on the market is important and simple QC methods for antibiotics and other essential drugs were developed for CE with cheap, simple, robust electrolyte systems and contactless conductivity detection [35-37].

MS-detection is regularly used for the identification for unknown impurities. In addition, Van Wijk *et al.* developed a CE-MS methodology for the separation and detection of potentially genotoxic alkylation compounds [38]. With electrokinetic stacking and the use of an internal standard, detection levels of as low as 10 mg/kg were obtained with repeatabilities of 6 – 8% RSD. Detection limits lower than 1 mg/kg were achievable with derivatization and full-scan mode (untargeted screening), which was at least as good as or better than full-scan LC-MS.

In order to be able to separate uncharged analytes, Terabe introduced micelles in the electrolyte solution [39]. Neutral analytes partition between the aqueous phase and the micellar phase and migrate part of the time with the aqueous phase (electroosmotic flow) and part of the time with the micellar velocity. Their net mobility depends on the partitioning, that is, the average time spent in the micelle and the average time spent migrating with the Electroosmotic flow (EOF). Charged analytes both have electrophoretic mobilities as well as partition into the micelles. In one of the earliest publications from industry, Johnson&Johnson people used micellar electrokinetic chromatography (MEKC) for the separation of amino acids and small peptides [40]. A validated stability-indicating method for diuretics hydrochlorothiazide and chlorothiazide was published by Merck [41]. By taking good care of the injection procedure and by frequent buffer vial changes, precision and reproducibility results of 1% RSD were achieved, combining HPLC reproducibility with CE efficiency. At AstraZeneca, a method for the analysis of cough tablets containing ibuprofen and codeine was developed and validated with help of multifactorial Design of Experiments (DOE) [42, 43]. The method proved suitable for assay of both active substances and the determination of related degradation products and impurities, and was submitted to the regulatory authorities.

Micro-emulsion electrokinetic chromatography (MEEKC) is a CE technique that can be applied over a large range of analytes. Charged as well as uncharged analytes can be separated over a wide range of lipophilicity. Interesting applications are the direct analysis of paracetamol suppositories [44] and poorly-soluble new chemical entities (NCEs) in liquid-filled capsules [45]. An aqueous solvent is needed to dissolve the gelatin capsules, whereas the hydrophobic drug needs a lipophilic environment. Micro-emulsions combine both. MEEKC can also be used as a platform method. At GlaxoSmithKline, a MEEKC method replaced at QC manufacturing a number of LC and spectroscopic methods for identity confirmation of drug substance in a range of product types and compositions such as immediate and sustained release tablets, creams, or capsules [46]. In order to be able to use MEEKC as a platform methodology, Harang at AstraZeneca studied the composition of the micro-emulsion and its effect on the separation in a multi-factorial design of experiments [47].

CZE, MEKC, MEEKC, chiral CE and affinity CE (ACE) are not only valuable tools in pharmaceutical development. This combination of techniques also offers added value in early drug discovery as high-throughput tools for physicochemical characterization such as pK_a and log P and protein binding studies of potential new drugs when very small amounts of the new drugs are available [48].

Enantiomeric Purity

Chiral analysis deserves special attention, as the strength of chiral CE is in its simplicity. In chiral CE, a chiral selector is added to the electrolyte solution. The enantiomers interact with the chiral selector. Separation is based on different equilibrium contacts of the enantiomers with the selector, different mobilities of the complexes or both [49-51]. No expensive columns are needed and screening for the right selector is an overnight exercise. Additionally, the separation can be optimized by varying the concentration of the selector in the electrolyte solution. Due to the high efficiencies achievable in CE, already a small difference in interaction with the chiral selector gives separation. For Chiral CE, industrial publications peaked in the period of 1993 – 2006. Chiral separation is usually accomplished by the use of a (derivatized) cyclodextrin as chiral selector. One of the first industrial applications was the chiral separation of clenbuterol and similar β-amino alcohols [52]. Advantage of the CE method compared to the LC method was that the drug substance could be directly dissolved in water, whereas LC needed extraction. Additionally, in the aqueous CE system UV detection at low UV-wavelength was possible, enhancing sensitivity with a factor > 10. Other benefits mentioned were lower running costs and reduced health & safety risks.

Wren proposed a simple model [53] for chiral CE that has served well in developing robust methods for chiral drugs. For example, the method for enantiomeric purity determination of ropivacaine in the pharmacopoeias was developed and studied based on the model and the method works well for both drug substance as well as drug product solutions for injection [54-56]. The method has proven very robust over the years and is still being used for the release of enantiomerically pure drug substance.

Wren's model assumes a 1:1 complex formation between enantiomer and chiral selector. Interaction with a stoichiometry of more than one selector with the enantiomer was observed in the case of a μ-opioid receptor binding tetrapeptide [57]. Other extensions considered more interactions, such as the charge state of the enantiomers. An enantiomer in the charges state will have a different interaction with the chiral selector than when it is neutral [58-62].

In the 1990s and at the turn of the century, several pharmaceutical companies had CE as a first choice for chiral separations. Many had special method development strategies, e.g., [51, 63-67]. The Pfizer strategy uses a mixture of sulfobutyl ether-β-cyclodextrin, γ-cyclodextrin and dimethyl-β-cyclodextrin in phosphate buffer pH 2.5. With this mixture, more than 90% of 26 randomly selected chiral drugs or intermediates were separated [63]. Merck screens different sulfated cyclodextrins for basic compounds [64]. The Janssen approach takes into account that enantioselectivity cannot be predicted and first screens for pH and then a Taguchi designed chiral selector screen is performed [65]. With this approach, basic, acidic and neutral analytes are covered. This screen is systematically applied at Janssen with satisfactory results. Boehringer Ingelheim also concluded that a universal method is not feasible and their approach for basic analytes comprises a screen of ten charged and uncharged cyclodextrins. Additionally they describe some measures for improving precision, such as swopping inlet and outlet vial between injections to prolong the lifetime of the buffer [66]. The approach described by Sänger – van de Griend reflects strategies within AstraZeneca and Solvay Pharmaceuticals [51, 67].

Chiral CE was also instrumental in proving the need for time- or velocity- corrected peak areas in CE, by applying this to a racemic mixture of two well-resolved enantiomers [68].

For small-molecule drugs, it is quite common to develop multi-purpose (or multi-attribute) methods. An example is the analysis of ximelagatran. One method is suitable for assay and purity and chiral purity analysis of both drug substance and drug product [69]. All related substances could be reliably quantified at the 0.05% corrected peak area level.

Counter Ions

There are many applications of drug counter ion analysis and although some say that chiral analysis are the most used applications in small-molecule pharmaceuticals, some find counter ion analysis to be more applied [5]. In our survey, about half the respondents replied that they use CE for the determination of ions or counterions. However, many of these methods were never published [5]. Additionally, there are several commercial kits available that fulfil the method requirements most of the times [70]. Certainly, there is a trend in using a wider spectrum of counter ions in drug salts [71]. As small cations and anions often lack a chromophore, applications typically use indirect-UV or contactless conductivity detection [70]. More recent applications are reviewed in [32, 72-74]. Altria described two methods for anionic and cationic drug counter ion determinations based on indirect UV detection [75]. The determination of sodium was evaluated in an intercompany study [76]. Five out of six companies obtained good levels of accuracy and precision. The sixth lab had issues, that were contributed to contamination of glassware or autosampler vials, demonstrating some of the general issues for analysis of these omnipresent analytes.

A more recent publication of a method with indirect-UV detection shows the simultaneous determination of the drug counterion chloride together with synthesis impurity acetate [77]. As acetate can also be the degradation product of sample prep, the authors took special care not to induce acetate formation during analysis. The paper carefully describes all issues and consequent choices in method development of a CE-indirect-UV assay.

Stålberg published an example of the determination of counterion impurity bromide in hydrochloride salts of a local anaesthetic [78]. Because bromide shows good absorptions at low-UV wavelengths, the amount of bromide was assayed with a direct CE-UV method. Accuracy and precision were well within the required limits, even without the use of an internal standard. The limits of detection and quantitation for the defined method were 0.36 μM and 1.2 μM, respectively, in a 6 mM drug hydrochloride salt, but could easily be pushed further down if needed, as bromide and chloride are well resolved. The reported LOQ was 10x lower than the required range for the determination [78].

Carbohydrates

There are currently many different ways of analyzing glycoproteins and CE is one of them. A recent review on glycan analysis by Zhang lists different approaches for intact proteins, glycopeptides, released glycans and monosaccharides [79]. CE plays a role at all these levels [80-82] and methods are still being updated and improved.

A simple methodology for mono- and oligosaccharide analysis is based on the work of Rovio [83]. At a high pH, such as in NaOH solutions or NaOH / phosphate buffers, enolate ions are formed that are negatively charged and absorb UV light at 265 nm. This methodology can be applied to samples from fermentation or cell cultures and analyses monosaccharides, disaccharides, aminoglycans and/or polyols [84, 85]. The alternative methodology uses CE with laser induced fluorescence detection (CE-LIF) after ATPS (9-Aminopyrene-1,4,6-trisulfonic acid) labeling. This methodology has been refined over the years, both the derivatization has been optimized and the process made easier and more robust with the use of magnetic beads [86]. The sample prep with magnetic beads simplifies the process, but also without the use of magnetic beads the procedure gives well reproducible results, as shown by an intercompany study [87]. Multiplexed CE systems such as commercialized for DNA-analysis can be used to speed up and reduce the total analysis time [88].

In order to get information on where glycan groups are attached to the proteins, the protein can be digested after which CE-MS analysis can map the peptides and give structural information on the glycans (e.g., [89, 90]).

Intact proteins can also be analyzed on the glycan structure. A simple approach is to determine the sialic acid heterogeneity. In use for many decades is the method for erythropoietin (EPO) [91, 92]. An interesting application of CZE at low pH for an intact highly sialylated protein was published by Amgen [93]. However, most attention for glycan characterization these days goes to CE-MS methodologies comparable to methods used for proteomics. With CE-MS under mild conditions, intact proteins are analyzed and glycan structures are derived from high-precision MS/MS data (e.g., [94-96])

Toppazzini wrote a comprehensive review on the analysis of polysaccharides with CE [97]. Another overview can be found in [98]. Although there have been some presentations on the analysis of polysaccharides, such as a MEKC method for the characterization of glycoconjugate vaccine [99], there are hardly any publications from the pharmaceutical industry on this topic.

Görög in a recent critical review on impurity profiling [100] underlined the importance of CE for the separation of heparin from oversulfated chondroitin and dermatan sulfate in the recent heparin crisis [101]. The CE method is capable of detecting oversulfated chondroitin sulfate and dermatan sulfate down to 0.05% and 0.5% (w/w), respectively [102].

Other Pharmaceutical Related Small Molecules

Besides drug-related assays and impurities, an area that deserves attention is the analysis of excipients. Zhu and Scriba reviewed recent publications on excipients such as artificial sweeteners and cyclodextrins [32].

In a publication in Farmazeutische Industrie in 2011, Solvias brought to the attention that there are quite large amounts of small molecules present in biopharmaceutical process samples and drug products and that many of these can be analyzed with CE [103]. Examples mentioned include reducing agent TCEP (Tris(2-carboxyethyl) phosphine), buffers like TRIS (Tris(hydroxymethyl)amino-methane), histidine and citrate, metal ions like magnesium and potassium, amino sugar acids like muramic acid and N-acetyl-muramic acid and additives like cyclodextrins.

Peptides

CE has always been in scope for the characterization of peptides and proteins. Organon in collaboration with the Eindhoven University of Technology used CZE to separate a modified adrenocorticotropic hormone peptide from its potential synthetic byproducts or metabolites [104]. From the pIs of the fragments and the estimated charge-to-size ratios, migration order could be predicted accurately. Multifactorial experimental design-based method development and validation was used for the determination of impurities of a tetrapeptide at AstraZeneca [105]. The same tetrapeptide resulted previously in extension of the chiral model [57]. Genentech demonstrated that CE is an orthogonal technique compared with HPLC for the separation of deamidated peptides and especially charged peptides were better resolved with CE compared to HPLC [106].

Peptide mapping for the evaluation of protein heterogeneity has been used since the early days of CE. Amgen used peptide mapping for the evaluation of glycopeptide heterogeneity of EPO. Their peptide map segregated non-glycosylated and glycosylated peptides in two different regions [107]. Merck separated five synthetic heptadecapeptides from the p-loop region of p21 Ras proteins with MEKC [108]. Larsson and Lutz from AstraZeneca explored tITP-CZE-MS (Transient isotachophoresis-CZE-MS) for sensitive analysis of peptides within DMPK and Bioanalysis [109]. Genentech used CE-MS for mapping the tryptic digest of therapeutic mAbs [110]. Indeed, the use of CE-MS for peptide mapping, amino acid sequence confirmation and glycosylation studies is becoming an established tool in protein characterization. An interlaboratory study showed that even with different equipment being used, the results of peptide mapping were comparable between laboratories [111].

Proteins

Capillary Zone Electrophoresis

Early on, people realized the power of CE for the separation of proteins. Intrinsically, the low diffusion coefficients of proteins result in high efficiencies in CE as long as adsorption of proteins to the fused silica wall is prevented [112, 113]. The first industrial paper on CZE for protein pharmaceuticals we traced is from Genentech. Frenz et al. characterized human growth hormone with a simple phosphate buffer pH 2.56 in a coated capillary [114]. Analysis of a fraction collected on a tryptic digest showed one peak for HPLC while two peaks were detected in CZE, demonstrating the added value of CZE.

Also NovoNordisk was early with exploring capillary electrophoresis for protein pharmaceuticals. A robust and precise stability-indicating method for the separation of human insulin and deamidation products in solution was developed [115]. In order to reduce interaction of protein with the fused-silica capillary wall while maintaining currents below the risk of excessive Joule heating, zwitterions were added to the sodium acetate electrolyte solution. The optimized electrolyte composition was 50 mM sodium acetate and 0.85 M N-Cyclohexyl-2-aminoethanesulfonic acid (CHES) in 10% acetonitrile, adjusted to pH 7.8 with triethylamine. The addition of acetonitrile both improved the plate numbers as well as the solubility of CHES. Additionally, the paper also demonstrates that the frequently suggested NaOH rinse between injections is not necessary and excellent migration time repeatability was obtained without basic wash.

Amgen set the basis of the glycoform separation method of erythropoietin in the pharmacopoeias [116], inspired by earlier work from Johnson&Johnson [117]. Resolution of the glycoforms was ultimately obtained by the electrolyte additives 1,4-diaminobutane and urea. Ultimately, CZE was taken up in the pharmacopoeial monographs for EPO [91, 92] and is nowadays being used by a considerable number of producers.

There are several early applications of CZE for the heterogeneity determination of monoclonal antibodies that use an 6-aminocaproic acid electrolyte solution, e.g., [118]. However, the use of CZE for charge heterogeneity really took off with the publication of Pfizer of a method based on high concentrations of 6-aminocaproic acid at pH 5.7 [16]. This method now serves as a platform method for mAbs with pIs ranging from 7.5 – 9.5. 6-Aminocaproic acid is a zwitterionic molecule with pKas of 4.43 and 10.75 [119]. At pH 5.7, the amine group is fully protonated, while about 94% of the carboxylic acid group is deprotonated, so the net charge of the zwitterion at pH 5.7 is close to zero. As a consequence, the conductivity of the electrolyte solution is very low and high concentrations can be used without resulting in excessive Joule heating. By using high electrolyte concentrations, in combination with addition of diluted hydroxypropylmethylcellulose (HPMC) and rinsing with hydrogen chloride (HCl) between injections, there is less protein adsorption and less electromigration dispersion [112, 120]. Adsorption of the proteins, especially of the basic species, to the capillary wall is further reduced by the addition of triethylenetetramine (TETA) to the electrolyte solution [16]. The selected pH is a balance between enhancing charge/size ratio differences, which are largest close to the pI, and having sufficient remaining mobility, that is, a pH not too close to the pI of the protein. At the pI, not only the net charge is zero with no electrophoretic mobility, also the protein is more prone to adsorption and solubility is lower [121]. Both the low electrophoretic mobility and adsorption will result in peak broadening. Reduced adsorption is also attributed to rinsing with HCl, which additionally suppresses the EOF [122-125].

Another frequently used methodology is based on lower concentration acetate buffers with TETA and based on the work of Shi [126]. Also here resolutions were remarkably affected by the preconditioning procedure.

An observation on the use of CZE methods for proteins such as mAbs is that generally companies spend very little time on method development. Literature examples are followed, and if that does not work for the protein at hand, CZE is discarded as an option. Unfortunately, too often time is not granted to try to understand why certain methodologies fail in certain situations, or why they can be successful in other cases. This is a pity. CZE is a powerful technology that can be wider applicable if some time is invested in understanding adsorption and ways to prevent it, as well as ways to enhance charge-based separations of proteins.

Capillary Gel Electrophoresis

In CE-SDS, proteins are denatured with SDS, similar as in SDS-PAGE. Separation is performed on a capillary filled with a gel buffer. The electroosmotic flow in these systems is virtually non-existent and when applying a negative field strength, SDS-denatured proteins migrate towards the anode and are separated by size. After experimenting with cross-linked gels, the current application uses physical, non-crosslinked gels. Very similar results and precisions were found when comparing SDS-PAGE and CE-SDS [11]. The most striking difference is that the use of a non-crosslinked dextran gel, such as in the CE-SDS kit [127], means that for proteins with side structures such as glycans, larger apparent molecular weights are found than the expected molecular weights. This has to do with the different degree of binding of SDS to these kind of proteins [128]. As a consequence, the charge-to-size ratios are altered and inaccurate molecular weights are estimated. Guttmann suggests an alternative way of estimating the accurate molecular weight by the Ferguson method with different gel dilutions [11]. The CE-SDS method is currently a platform method for the characterization of mAbs and although the apparent molecular weight of the glycosylated protein is larger than the actual molecular weight, the methodology serves very well as a general purity test [12].

Genentech published a well-written paper describing the fluorescence labeling procedure they use in an analytical Quality by Design way [129]. When working with fluorescence detection, baseline disturbances that are characteristic for the gel and that can impede peak integration with UV detection, are no longer a problem.

In an opinion article in the Analytical Scientist [130], Winkhaus from Roche underlines the importance of higher throughput than the traditional capillary CE-SDS approach can give. They implemented 45 times faster chip-sized CE-SDS (μCE-SDS) and where they assumed a loss of resolution, actually resolution was superior in some cases. Additional advantages where lower failure rates, although a disadvantage was reduced reproducibility. Combined with automated sample preparation, the remaining bottleneck is data analysis. Others that have implemented μCE-SDS for GMP applications, both for release as well as for stability studies, are GSK [131] and Merck [132]. Amgen implemented automated sample prep to improve precision [133].

Besides CE-SDS on chip, there are new developments on the instrumental side, as other vendors have introduced dedicated CE-SDS instrumentation. As all these different formats of CE-SDS use different patented gel buffers, we cannot evaluate the pros and cons of these gel buffers in a scientific way. Consequently, if companies want to replace the existing methodology with the methodology from another vendor, extensive bridging studies for each therapeutic molecule are required.

Although the original CE-SDS application was developed for Immunoglobulin G1 (IgG1) monoclonal antibodies, the kit has successfully been used for other modalities such as bispecific antibodies [134], antibody-drug conjugates [135], fusion proteins [133] or virus proteins [136]. For deglycosylated and reduced influenza proteins, the use of diluted gel buffer gave better separations, despite these proteins being similar in size as the reduced mAb chains [136].

CE-SDS being well established in industry, the need for identification of the detected proteins increases. Although different strategies include reduction, deglycosylation, digesting and other treatment of proteins to induce controlled fragmentation, there is a need for further analysis. An interesting development is the coupling of CE-SDS to mass spectrometry. Due to the high amounts of SDS, direct coupling is not feasible. The group of Neusüß in collaboration with Roche is developing an automated system where the peak is cut from the CE-SDS run, desalted and introduced to the MS with a CE-SDS–CZE–MS coupling [137].

Capillary IsoElectric Focusing

In capillary isoelectric focusing, proteins are separated by charge. The capillary is filled with a pH-gradient, in which the proteins migrate to their pIs, where the net charge is zero. With no net charge, they no longer have an electrophoretic mobility and remain at the position where the pH equals the pI. Early on professor Hjertén [138, 139] and Thormann [140, 141] realized the importance of cIEF both for focusing diluted samples and for the high resolutions that can be obtained. Capillary IEF is a spatial separation, and after the focusing the protein bands need to be detected. This can either be done by mobilization of the proteins so that they migrate through the detector [142], or by whole-capillary imaging as in dedicated icIEF instruments [143, 144].

Recently, a new imaging instrument came on the market that can detect native fluorescence [145]. Where UV-detection is limited to 280 nm because of absorption of ampholyte at lower wavelengths, native fluorescence results in cleaner baselines. This makes it easier to use ampholytes that usually are avoided for UV detection, such as Servalyte, an ampholyte that generally gives better resolution for fusion proteins. Native fluorescence is said to give 2 – 5 times higher sensitivity than UV-detection at 280 nm [145]. For many years, people have been using the traditional ampholytes from gel IEF to form pH gradients in the capillary. However, recently there are some efforts to start producing UV-transparent ampholytes [146, 147]. Another new development is the coupling of (i)cIEF to mass spectrometry. A review of Neusüß et al. gives a good indication of the advantages and challenges for different coupling strategies [148]. A first example from industry applying icIEF-MS to separate charge variants of mAbs with online time-of-flight mass spectrometry (TOF-MS) detection and characterization was recently published [149].

Affinity and Immunoaffinity CE

The keyword "affinity" means that affinity CE (ACE) covers a large portion of the publications on CE. This includes publications with any kind of chromatographic affinity interaction, such as MEKC, MEECK, chiral CE or CEC with molecular imprinted polymers. If we observe these groups as a whole, most of the papers were published in the period 1993 – 2008. If we narrow ACE down to include papers where the focus is on studying the interaction between molecules (such as drug-protein, protein-metal ion, receptor-ligand, antibody-antigen interactions) or where (immune)affinity-based methodologies were used for selective sample preparation, the number of publications from industry reduce to about 33 over the whole time period. This is comparable with the amount of industrial publications on chip applications. Most of these publications were on the turn of the century (16/33 in the period 1998 – 2002). In general, there are two ways of determining affinity interactions in a CE system. If the ligand-receptor kinetics is slow, the sample is pre-equilibrated (or incubated) before the separation and separation is based on different mobilities between the free ligand or receptor and the complex [150]. Rustandi et al. used this principle to characterize the binding properties of a therapeutic monoclonal antibody to its target antigen and separated the complex from the free with icIEF [151]. If the complexation kinetics are fast, one of the components can be added to the electrolyte solutions and by varying the concentration, and thereby influencing the mobility, complexation constants can be determined. A review explaining the principles was written by Heegaard et al. [152]. ACE is a very powerful technique, but it seems more used in the academic domain than in industrial applications, although affinity and immunoaffinity applications of CE could add specificity to many potency tests.

Important work on affinity and immunoaffinity on-line and/or in-capillary sample preparation and stacking was done by Norberto Guzman [153]. The paper describes several strategies to use immunoaffinity approaches to increase sensitivity. Unfortunately, this visionary work of Guzman is still hardly applied. Industrial application of CE could well benefit from commercialization of these type of solutions.

Other Modes

There are a few more developments on the analysis of proteins to mention. On the one hand, more types of application instruments come available, such as capillary western blotting [145, 154]. On the other hand, the capillary electrophoresis instrument is being used in a clever way to characterize protein interactions. For instance, Taylor dispersion can be used to determine diffusion coefficients, protein size or protein aggregation [155-158], but also to determine affinity between probe and ligand in so-called flow-induced dispersion analysis (FIDA) [159, 160]. Proteins can be quantified in complex matrices by binding to a specific ligand with high affinity. Because of the binding, the apparent size of the ligand will change and hence its diffusivity will change. These types of applications are currently finding their way towards the pharmaceutical industry.

Nucleotides

DNA analysis is perhaps the most used application of CE, as the CSI tv-series show us every night. Many scientists focused on nucleotide analysis by CE in all its aspects in the 1990s, and in this period we also see several papers from industry. At Novartis, transient isotachophoresis-capillary gel electrophoresis (tITP-CGE) was used to analyze modified nucleotides, phosphorothioates, that were being investigated as potential therapeutic [161]. The paper describes the tITP development well. GlaxoWellcome used CE to analyze a complex of plasmid DNA with a liposome that was being developed as a treatment for cystic fibrosis [162]. Linear, supercoiled and open-circle conformers of plasmid DNA were separated from each other and from the complex with liposome and the method was used as a quality control and stability determination. Pfizer used CGE-LIF for separating plasmid DNA topoisomers and a dimer form [163]. Merck optimized a CGE method for the determination of viral dsRNA fragments with relatively little sample prep needed, even for complex sample matrices [164]. AstraZeneca applied on-column electroextraction to improve sensitivity of oligonucleotide analysis with CGE from human plasma [165]. Solvias used CGE for the separation of therapeutic single- and double-stranded oligonucleotides and optimized existing methodologies to fit pharmaceutical analysis [166]. Recently, different companies have started looking into protein – nucleotide complexes with CE [167].

Viruses and Bacteria

Amstrong et al. published some excellent reviews in 2003 and 2004 about the separation of microorganisms including viruses, bacteria, cells, and fungi by capillary electrophoresis [168, 169]. In 2007, Kenndler and Kremser issued a good review describing capillary electrophoresis for viruses, subviral particles and virus complexes giving some cIEF, CZE and affinity CE examples [170].

Viruses

Hjertén et al. was the first to describe the CE analysis of a virus in 1987 [139]. Tobacco Mosaic Virus (TMV) was analyzed in a bare-fused silica capillary using TRIS-HCl at pH 7.5. In the 1990s, CE was elaborately studied for the analysis and characterization of the human rhinovirus. Kenndler et al. described the pI determination of the rhinovirus by cIEF [171], the interaction between neutralizing monoclonal antibodies and human rhinovirus (serotype HRV2 and HRV14) by ACE [172], and the challenges of the peak identification of HRV by CZE [173].

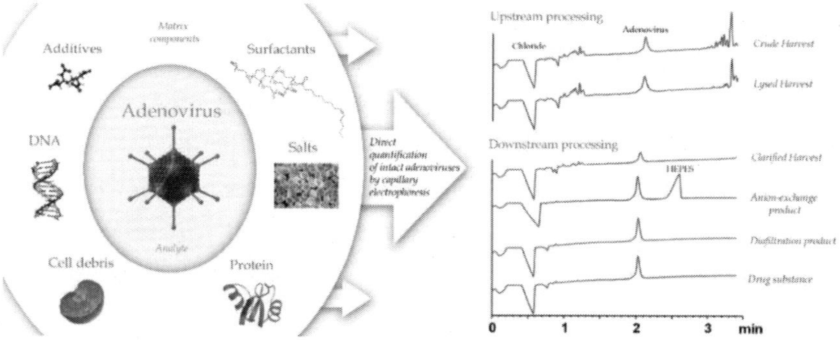

Figure 2. Capillary zone electrophoresis of Adenovirus type 26 samples from upstream processing and downstream processing: benzonase-treated crude harvest, benzonase-treated lysed harvest, clarified harvest, anion exchange product, diafiltration product, and drug substance. Reprinted with permission from Ref. [184]. Copyright (2017) Elsevier.

Most recently, Oita and Halewyck et al. exhaustively studied the poliovirus, poliovirions and subviral particles in a series of publications [174-176]. CZE was used to separate the poliovirus from cell extracts and the analysis was extensively optimized by a Placket-Burman design. In addition, the impact of the sample matrix composition and sample stacking on the signal enhancement of the poliovirus was studied [177-179]. ACE was used to evaluate the complex formation between the poliovirus and nanobodies [180].

Bettonville et al. described the development of an CZE method for the analysis of intact virus-like particles of human papillomavirus [181, 182].

In 2000 Mann et al. published the CZE analysis of recombinant adenovirus on polyvinyl alcohol-coated (PVA) capillaries by using sodium phosphate as BGE [183]. Geurink et al. published a novel CZE method for the quantification of intact adenovirus particles throughout upstream and downstream processing. The adenovirus was separated from salts, excipients, cell debris, DNA and protein in a PVA-coated capillary using TRIS-Tricine at pH 7.7, see Figure 2 [184].

As described before, CGE has extensively been used as platform method for antibody characterization. A recent observed trend is the use of CGE for the analysis of viral proteins from viruses and virus-like particles, such as for the in-process testing of recombinant Hepatitis G viral proteins [185], glycan analysis of influenza A virus [186], analysis of the VP7 glycoprotein of rotavirus [187], the determination of rotavirus-like particles [188], the quantification of influenza proteins from influenza viruses and virus-like particles (Figure 3) [136], and the characterization of Western, Eastern, and Venezuelan equine encephalitis (WEVEE) virus like particles (VLPs) [189].

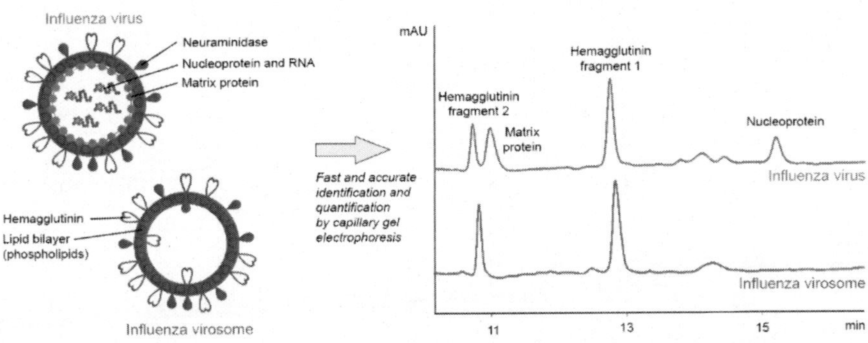

Figure 3. CGE analysis of the influenza virus and influenza virosome (virus-like particles). Reprinted with permission from ref. [136]. Copyright (2015) Elsevier.

Bacteria

One of the first publications describing the analysis of bacteria by capillary electrophoresis was in 1987 when Hjertén et al. published the analysis of the Lactobacillus casei bacteria by CZE using a fused silica capillary [139, 190]. Kremser et al. gave a nice overview about the use of

CE for bacteria in their review paper from 2004 [170]. Rodriguez et al. [169] and Desai [168] et al. also described multiple CE applications for bacteria. Amstrong and Lantz et al. [191] described the optimization of a capillary electrophoresis method for single-cell detection of bacteria and fungi to test the microfluidic sterility. They published a *Stimuli* paper in Pharmacopeial Forum to propose CE for sterility testing as an alternative to time-consuming inoculation or PCR methodologies [192].

CZE, cIEF and matrix-assisted laser desorption/ionization time-of-flight (MALDI-TOF) mass spectrometry have been used for the characterization of Dickeya and Pectobacterium species. The three methods were successfully used to classify the species. Breadmore et al. published an innovative isotachophoretic (ITP) method in combination with inline fluorescence in situ hybridization (FISH) for the detection of intact bacterial cells. The methodology is specific for bacterial strains using sequence-specific probes and seems a promising alternative for classical approaches used to detect bacteria.

Platform Methods

The term "platform methods" is frequently used within (bio)pharmaceutical analysis these days. Over the years, different names have been used for similar strategies, *viz.* method development strategy, method development starting point, generic method and recently platform method. None of these approaches are one-size-fits-all. After all, it would be striking if the same physicochemical properties that determine the efficacy and safety of a drug molecule are unique for its functionality and uniform for separation science. However essentially, based on the physical-chemical properties and/or class of the molecule, a strategy is followed that ultimately comes back to applying sound fundamental knowledge about the capillary electrophoresis techniques and understanding the potentials and the limitations. Platform or generic methods fail if applied without understanding the fundamentals. Several of these platform methodologies are discussed in this chapter. It is however, up to the reader to assure (if needed with help of an expert) if a platform methodology is applicable and what the limitations are for the molecule at hand.

ANALYTICAL QUALITY BY DESIGN

A recent trend is the application of Analytical Quality by Design (AQbD) for analytical method development in the pharmaceutical industry. The first publications describing the AQbD principles raise from about 2007. Between 2007 and 2011, 19 papers were published on QbD in combination with analytical method development. This number increased significantly between 2012 and 2018 with a total of 120 publications. The numbers of publications on AQbD and CE also increased the last years with no publications until 2011 and 19 publications between 2012 and 2018.

What is AQbD?

Analytical quality by design is a systematic approach used for method development to assure that the right method is developed for the right purpose and that the sources of variation are known and under control during its lifecycle, see Figure 4. The aim of an analytical method is described in an analytical target profile (ATP). This profile captures the objective of the test and the quality requirements for the reportable result. The quality requirements can be split in method requirements (e.g., precision and sensitivity of the method) and in business requirements (e.g., method needs to be validated in QC or the method should be delivered within 3 months). Since it is often impossible to meet all requirements from the ATP during method development, the requirements are put in order of priority.

First, the characteristics of the analytical technologies from pharmacopoeia, in-house expertise or literature are compared with and scored towards the requirements from the ATP. The selected technique or techniques should meet most of the requirements or should have the least risks or uncertainty to meet the requirements. Before method development is started, the (potentially) critical method parameters are defined for the technique of choice by using a criticality assessment and/or risk assessment based on scientific knowledge. Typically, the criticality assessment and risk assessment are repeated throughout the lifecycle of the method as soon as (practical) knowledge is gained. In the end, the risks from all critical method parameters should be mitigated and be in control.

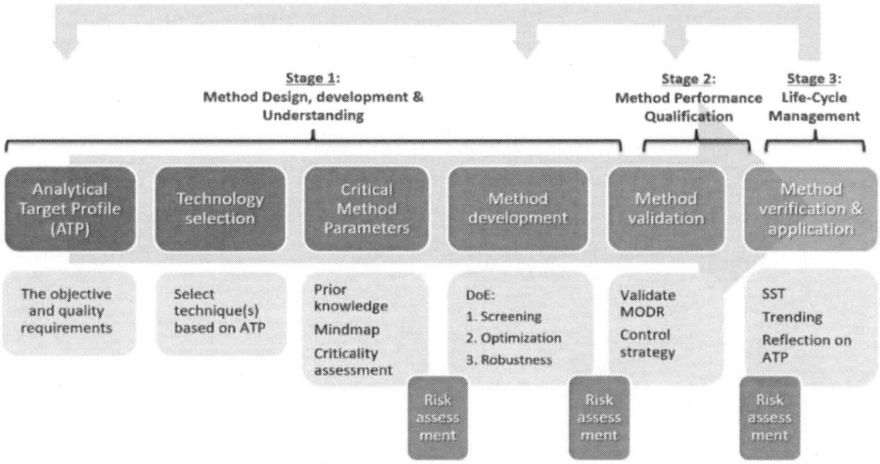

Figure 4. The analytical quality by design process.

All decisions made during method development need to be reflected on the ATP requirements. Method development itself is focused on the critical method parameters with the highest risk or greatest risk. Preferably, design of experiments (DOE) is used for the experiments instead of the traditional single-variable approach. Three types of designs can be differentiated; 1) screening DOE, 2) optimization DOE and 3) robustness DOE. The screening design is used in case the number of critical method parameter is still very large (> 10) after the criticality and/or risk assessment. The screening design focuses on the main effects of the parameters and is used as a filter tool to select the parameters with the largest effect that should be optimized in the optimization DOE.

The optimization DOE is used to find optimal settings or ranges for the critical method parameters (typically < 5). In this kind of design not only the main effects are assessed but also the interaction and quadratic effects of the parameters. The final result is typically a design space were the behavior of parameters and the combination of parameters can be predicted by a statistical model.

The robustness design is used in case all critical method parameters are optimized and you want to know the robustness of the selected optima. Typically, the parameters are tested with slight variations (e.g., ± 5%), representative for the variation expected when the method is in use after implementation. After the robustness experiment you should know the range of each parameter that would result in an acceptable result (e.g., temperature of the capillary cassette: robust range: 14 – 16°C, set point: 15°C).

The last steps of the analytical quality by design process include method validation and method life-cycle management. The method validation is commonly done according ICH guideline Q2-R1, but in literature it also described that the whole method operable design region (MODR) is preferably being validated [193, 194]. The method life-cycle is maybe the most important step of the whole process. It is important to note that the method validation is not showing the true performance of the developed method. Only during the life-cycle the method will demonstrate its performance. The method will be transferred to other labs, will be operated by different operators, different suppliers will be used for chemicals and materials, the products being analyzed change etc. It is important to assure during the life-cycle that the method is still performing as it was intended to. It is a good habit to check regularly if the method performance is still meeting the requirements as described in the analytical target profile (ATP). A system suitability test should be in place to control the critical method parameters of the method and the system suitability controls and assay controls should be trended and monitored.

Will AQbD Be Expected from Us in the Future?

A *Stimuli* article was published that describes the framework of a proposed USP chapter "<1220> The Analytical Procedure Lifecycle" [195].

The purpose of the proposed chapter is to define concepts that can be used to control analytical methods during their entire lifetime and to be consistent with ICH (International Council for Harmonisation) guidelines Q8-R2, Q9, Q10, and Q11. The current thinking of the USP Validation and Validation Expert Panel is very much in line with the Analytical Quality Design Approach described in literature [194] and it is expected that this approach will be recommended by the USP in the future [204, 211].

CE Applications with AQbD

Some good examples describing (aspects of) application of AQbD in combination with CE are the development of a CE assay for the determination of dextromepromazine and levomepromazine sulfoxide [196], the determination of the chiral purity of dextromethorphan by CZE [197], the quantification of intact adenovirus particles by CZE [184], the determination of captopril, hydrochlorothiazide and their impurities by MEKC [198], the quantification of ciprofloxacin by CE-C^4D [35], the development of a capillary electrophoresis method for the analysis of zolmitriptan and its impurities [199], the development of a capillary electrophoresis method for the analysis of almotriptan and its impurities [200] and the quantitative impurity analysis of monoclonal antibody size heterogeneity by CE-LIF [129].

Expectations for Future Use

The described applications of CE in industry illustrate that CE is widely applicable for pharmaceutical analysis. Looking at the trends in the use of CE in industry, there are a few obvious predictions for future use. The urgent need for identification of unknown peaks and the new impulses on CE-MS interface development means that we will see an increase in the use of CE-MS for pharmaceutical analysis and especially for protein characterization and biomarker exploration. If we may add some wishes to the expectations, it would be that new designs of coupling techniques of CE to MS take better into account the strengths from CE and from MS, rather than adapting existing interfaces and adapting them to make CE-MS possible. Ideally, we wish for CE-friendly MS interfaces, rather than MS-friendly electrolyte solutions. We have seen that the electrolyte composition plays a larger role than setting the pH and thus the charge on the analyte [29]. Especially for advanced protein separations we will need to have free hands in selection of the electrolyte composition. Also knowledge for reducing and prevention of adsorption is needed that is better compiled, more fundamental and less application-dependent.

We currently see two diverging tracks for (protein) pharmaceutical analysis. One the one hand, QC laboratories and management like dedicated application instrumentation with kit chemistry to make the analysis easy to handle and very reproducible. On the other hand, kits with an unknown composition make AQbD and risk control difficult and additionally it is hardly likely that proteins with unique therapeutic properties will behave the same physical-chemically. So for in-depth analysis, method optimization for each specific protein will be needed. In our minds, after getting over the initial threshold, this will be easier for CZE methodology than for slow and unspecific separation methodologies as CGE. The CZE methodologies can well serve multi-attribute purposes.

Since the 1990s, we have been expecting more chip-based applications, and this finally seems to take a flight with a broader range of applications than previously. Still even here we would like these chip platforms to be more flexible so that separation conditions can be adopted for each sample and analyte. Interesting will be whether 3D-printing technology will play a role in this. There has been an increase in clinical application development in this area, such as personalized medical devices, at-patient analysis and even in-body monitoring. Here the trend is out of the lab (in)to the patient. Translating this to the pharmaceutical process, this means process analytical technology (PAT) in the form of in-process or in-line analysis in bioreactors and ultimately preparing the way for real-time release. The diversity of the CE technique is well suited for this, especially as CE is less sensitive to dirty matrices and requires less sample prep than other separation technologies. Multi-channel devices could improve throughput or sensitivity.

If we may dream a little further, all these new instrumental developments create the opportunity to improve injection precision and detection sensitivity. Injection should be able from any form of vial or well plate. Also, as CE is fundamentally bi-directional, we would like the possibility to inject the sample in the middle of the capillary/separation channel and detect both at the cathodic as well as the anodic end, preferably with multiple detection principles simultaneously. At least contactless conductivity detection, being cheap and non-interfering, should be built in each device at multiple points along the capillary. As for detection, we would love to see other detection principles, such as N- or P-specific detection, but also an extension of what is currently possible, such as multi-wavelength fluorescence detection.

All the data generated require more advanced data management software than is currently standard on CE equipment. We need software that not only is better in integrating typical CE peaks, but at the same time can convert time axis to mobility axis and relative mobility axis in a convenient way.

Further implementation and expansion of the CE techniques in industrial pharmaceutical analysis, not as an aim in itself but to serve the patients better, is only possible if we as the CE community keep sharing good practice and good solutions. Not only at national and international meetings, but even through scientific publications and the submission of robust methods to the pharmacopoeias, with system suitability requirements based on AQbD. If we continue doing this, CE will have a bright future in the pharmaceutical and biotech industry.

REFERENCES

[1] Tiselius, Arne WK. (1948). Electrophoresis and adsorption analysis as aids in investigations of large molecular weight substances and their breakdown products. *Nobel Lecture,* December 13, 1948.

[2] Hjertén, Stellan. (1967). Free zone electrophoresis. *Chromatographic reviews* 9 (2):122-219.

[3] Jorgenson, JW, and K DeArman Lukacs. (1981). Zone electrophoresis in open-tubular glass capillaries: Preliminary data on performance. *Journal of Separation Science* 4 (5):230-231.

[4] Muijselaar, Pim G. (2008). 7 Overview of current regulatory guidance. *Separation Science and Technology* 9:145-169.

[5] Sänger-van de Griend, Cari E. (2018). The application of capillary electrophoresis in the (bio)pharmaceutical industry. In *Flow and Capillary Electrophoretic Analysis*, edited by M Trojanowicz P Koscielniak. Nova Science Publishers.

[6] Sänger-van de Griend, Cari. (2012). Revival of Capillary Electrophoretic Techniques in the Pharmaceutical Industry. *LC GC North America* 30 (11):954-971.

[7] Kennett, Sarah. *CE for QC – A reviewer's perspective.* Presented at CEPharm, Seattle, WA.

[8] Guo, Amy, Gary Camblin, Mei Han, Charlie Meert, and SungAe Park. (2008). 14 Role of CE in biopharmaceutical development and quality control. *Separation Science and Technology* 9:357-532.

[9] Kennett, Sarah, Hermann Wätzig, Cari Sänger – van de Griend, Brian Nunnally, Kai-Jorrit Maul, Nadine Ritter, Richard Rustandi. (2015). *Invalid rate workshop. CEPharm*, San Diego, CA.

[10] Schenerman, MA, and SH Bowen. (2001). Optimization, validation, and use of capillary gel electrophoresis for quality control testing of Synagis®, a monoclonal antibody. In *CE in Biotechnology: Practical Applications for Protein and Peptide Analyses*, 66-74. Springer.

[11] Guttman, András. (1996). Capillary sodium dodecyl sulfate-gel electrophoresis of proteins. *Electrophoresis* 17 (8):1333-1341.

[12] Chen, AB, and E Canova-Davis. (2001). Capillary electrophoresis in the development of recombinant protein biopharmaceuticals. In *CE in Biotechnology: Practical Applications for Protein and Peptide Analyses*, 7-17. Springer.

[13] "*USP General Chapter 129 Analytical Procedures for Recombinant Therapeutic Monoclonal Antibodies*."

[14] Salas-Solano, Oscar, Kunnel Babu, SungAe Suhr Park, Xinfeng Zhang, Li Zhang, Zoran Sosic, Boris Boumajny, Ming Zeng, Kuang-Chuan Cheng, and Angelia Reed-Bogan. (2011). Intercompany study to evaluate the robustness of capillary isoelectric focusing technology for the analysis of monoclonal antibodies. *Chromatographia* 73 (11-12):1137-1144.

[15] Salas-Solano, Oscar, Babu Kennel, SungAe Suhr Park, Kelly Roby, Zoran Sosic, Boris Boumajny, Sarah Free, Angelia Reed-Bogan, David Michels, and Will McElroy. (2012). Robustness of iCIEF methodology for the analysis of monoclonal antibodies: an interlaboratory study. *Journal of separation science* 35 (22):3124-3129.

[16] He, Yan, Colleen Isele, Weiying Hou, and Margaret Ruesch. (2011). Rapid analysis of charge variants of monoclonal antibodies with capillary zone electrophoresis in dynamically coated fused-silica capillary. *Journal of separation science* 34 (5):548-555.

[17] Moritz, Bernd, Volker Schnaible, Steffen Kiessig, Andrea Heyne, Markus Wild, Christof Finkler, Stefan Christians, Kerstin Mueller, Li Zhang, and Kenji Furuya. (2015). Evaluation of capillary zone electrophoresis for charge heterogeneity testing of monoclonal antibodies. *Journal of Chromatography B* 983:101-110.

[18] CASSS. "*CE in the biotechnology and pharmaceutical industries on the practical application for the analysis of proteins, nucleotides and small molecules.*" www.casss.org.

[19] Altria, Kevin D. (2013). *Analysis of pharmaceuticals by capillary electrophoresis*. Vol. 2: Springer Science & Business Media.

[20] Sänger-van de Griend, Cari E. (2008). 6 General considerations to improve performance of CE methods. *Separation Science and Technology* 9:123-144.

[21] Altria, Kevin. (2007). *Buffer Preparation—Hints, Tips and Common Errors.*

[22] Altria, KD. "*CE Currents series in LCGC.*" Chromatography Online.com.

[23] Griend, CE. Sänger-van de. *CE Solutions series, Separation Science.*

[24] Michels, David A, Oscar Salas-Solano, and Chantal Felten. (2011). Imaged capillary isoelectric focusing for charge-variant analysis of biopharmaceuticals. *BioProcess Int* 9 (10):48-54.

[25] Felten, Chantal, and Oscar Salas Solano. (2010). Capillary Electrophoresis in Quality Control: PART I: Application for Therapeutic Proteins. *Beckman Coulter*:1-8.

[26] Felten, Chantal, and Oscar Salas Solano. (2011). Capillary Electrophoresis in Quality Control: PART II: CE-SDS: Method Development and Robustness. *Beckman Coulter*.

[27] Schmitt-Kopplin, Ph, F Menzinger, D Freitag, and A Kettrup. (2001). Improving the use of CE in a chromatographer's world. *LC GC EUROPE* 14 (7):384-388.

[28] Griend, MC Breadmore and CE Sänger – van de. (2014). In-Capillary Sample Concentration in CE. *LCGC*:171-186.

[29] Schepdael, CE Sänger – van de Griend and A Van. (2018). Chapter 10: Method Development and Validation of Capillary Electromigration Methods. In *Capillary Electromigration Separation Methods*, edited by CF Poole. Elsevier.

[30] Shapiro, Marjorie. (2018). *An FDA Perspective on the Implementation of State-of-the-Art Analytical Methods for the Development of Therapeutic Proteins*, presented at WCBP, Washington, DC.

[31] McKinsey&Company. https://www.mckinsey.com/industries/pharmaceuticals-and-medical-products/our-insights/rapid-growth-in-biopharma.

[32] Zhu, Qingfu, and Gerhard KE Scriba. (2017). Analysis of small molecule drugs, excipients and counter ions in pharmaceuticals by capillary electromigration methods–recent developments. *Journal of pharmaceutical and biomedical analysis*.

[33] Altria, KD, and NW Smith. (1991). Pharmaceutical analysis by capillary zone electrophoresis and micellar electrokinetic capillary chromatography. *Journal of Chromatography A* 538 (2):506-509.

[34] Wynia, GS, G Windhorst, PC Post, and FA Maris. (1997). Development and validation of a capillary electrophoresis method within a pharmaceutical quality control environment and comparison

with high-performance liquid chromatography. *Journal of Chromatography A* 773 (1-2):339-350.

[35] Paul, Prasanta, Christophe Van Laeken, Cari Sänger-van de Griend, Erwin Adams, and Ann Van Schepdael. (2016). CE-C4D method development and validation for the assay of ciprofloxacin. *Journal of pharmaceutical and biomedical analysis* 129:1-8.

[36] Paul, Prasanta, Cari Sänger-van de Griend, Erwin Adams, and Ann Van Schepdael. (2018). A simple, low-cost and robust capillary zone electrophoresis Method with capacitively coupled contactless conductivity detection for the routine determination of four selected penicillins in Money-constrained laboratories. *Electrophoresis*.

[37] Paul, Prasanta, Tom Duchateau, Cari Griend, Erwin Adams, and Ann Schepdael. (2017). Capillary electrophoresis with capacitively coupled contactless conductivity detection method development and validation for the determination of azithromycin, clarithromycin and clindamycin. *Journal of separation science*.

[38] Van Wijk, AM, HAG Niederländer, MD Van Ogten, and GJ De Jong. (2015). Sensitive CE–MS analysis of potentially genotoxic alkylation compounds using derivatization and electrokinetic injection. *Analytica chimica acta* 874:75-83.

[39] Terabe, Shigeru, Koji Otsuka, Kunimichi Ichikawa, Akihiro Tsuchiya, and Teiichi Ando. (1984). Electrokinetic separations with micellar solutions and open-tubular capillaries. *Analytical Chemistry* 56 (1):111-113.

[40] Tran, An D, Timothy Blanc, and Eric J Leoplod. (1990). Free solution capillary electrophoresis and micellar electrokinetic resolution of amino acid enantiomers and peptide isomers with L-and D-Marfey's reagents. *Journal of Chromatography A* 516 (1):241-249.

[41] Thomas, BR, XG Fang, X Chen, RJ Tyrrell, and S Ghodbane. (1994). Validated micellar electrokinetic capillary chromatography method for quality control of the drug substances hydrochlorothiazide and chlorothiazide. *Journal of Chromatography B: Biomedical Sciences and Applications* 657 (2):383-394.

[42] Persson-Stubberud, Karin, and Ove Åström. (1998). Separation of ibuprofen, codeine phosphate, their degradation products and impurities by capillary electrophoresis: I. Method development and optimization with fractional factorial design. *Journal of Chromatography A* 798 (1-2):307-314.

[43] Stubberud, Karin Persson, and Ove Åström. (1998). Separation of ibuprofen, codeine phosphate, their degradation products and impurities by capillary electrophoresis: II. Validation. *Journal of Chromatography A* 826 (1):95-102.

[44] McEvoy, Eamon, Sheila Donegan, Joe Power, and Kevin Altria. (2008). Application of MELC and MEEKC for the analysis of paracetamol and related impurities in suppositories. *Chromatographia* 68 (1-2):49-56.

[45] van Tricht, E, C Slooten, L van den Bos, R Maasman, C Sänger – van de Griend. (2009). Micro-emulsion electrokinetic chromatography: investigation of the applicability on Solvay drug products. presented at *CEPharm,* Boston, MA

[46] Altria, Kevin D, Nigel R Barker, Mairwen Hayworth, and Alex D Henderson. (2011). Use of a single capillary electrophoresis method as a routine identity test for batch release of a range of pharmaceutical products. *Chromatographia* 74 (11-12):783-787.

[47] Harang, Valérie, Jessica Eriksson, Cari E Sänger-van de Griend, Sven P Jacobsson, and Douglas Westerlund. (2004). Microemulsion electrokinetic chromatography of drugs varying in charge and hydrophobicity: I. Impact of parameters on separation performance evaluated by multiple linear regression models. *Electrophoresis* 25 (1):80-93.

[48] Wan, Hong, and Richard A Thompson. (2005). Capillary electrophoresis technologies for screening in drug discovery. *Drug Discovery Today: Technologies* 2 (2):171-178.

[49] Wren, Stephen. (2013). *The separation of enantiomers by capillary electrophoresis*. Vol. 6: Springer Science & Business Media.

[50] Chankvetadze, Bezhan. (1997). *Capillary electrophoresis in chiral analysis*: John Wiley & Sons.

[51] Sänger-Van De Griend, Cari. (1999). *Enantiomeric separations by capillary electrophoresis in pharmaceutical analysis*: Acta Universitatis Upsaliensis.

[52] Altria, KD, DM Goodall, and MM Rogan. (1992). Chiral separation of β-amino alcohols by capillary electrophoresis using cyclodextrins as buffer additives. I. Effect of varying operating parameters. *Chromatographia* 34 (1-2):19-24.

[53] Wren, Stephen AC, and Raymond C Rowe. (1992). Theoretical aspects of chiral separation in capillary electrophoresis: I. Initial evaluation of a model. *Journal of Chromatography A* 603 (1-2):235-241.

[54] Sänger-Van de Griend, CE, and K Gröningsson. (1996). Validation of a capillary electrophoresis method for the enantiomeric purity testing of ropivacaine, a new local anaesthetic compound. *Journal of pharmaceutical and biomedical analysis* 14 (3):295-304.

[55] Sänger-Van de Griend, CE, K Gröningsson, and Douglas Westerlund. (1996). Chiral separation of local anaesthetics with capillary electrophoresis. *Chromatographia* 42 (5-6):263-268.

[56] Sänger-van De Griend, CE, H Wahlström, K Gröningsson, and M Widahl-Näsman. (1997). A chiral capillary electrophoresis method for ropivacaine hydrochloride in pharmaceutical formulations: validation and comparison with chiral liquid chromatography. *Journal of pharmaceutical and biomedical analysis* 15 (8):1051-1061.

[57] Sänger-Van de Griend, CE, K Gröningsson, and T Arvidsson. (1997). Enantiomeric separation of a tetrapeptide with cyclodextrin extension of the model for chiral capillary electrophoresis by complex formation of one enantiomer molecule with more than one chiral selector molecules. *Journal of Chromatography A* 782 (2):271-279.

[58] Rawjee, Yasir Y, Daniel U Staerk, and Gyula Vigh. (1993). Capillary electrophoretic chiral separations with cyclodextrin additives: I. Acids: Chiral selectivity as a function of pH and the concentration of β-cyclodextrin for fenoprofen and ibuprofen. *Journal of Chromatography A* 635 (2):291-306.

[59] Rawjee, Yasir Y, Robert L Williams, and Gyula Vigh. (1993). Capillary electrophoretic chiral separations using β-cyclodextrin as resolving agent: II. Bases: Chiral selectivity as a function of pH and the concentration of β-cyclodextrin. *Journal of Chromatography A* 652 (1):233-245.

[60] Rawjee, Yasir Y, and Gyula Vigh. (1994). A Peak Resolution Model for the Capillary Electrophoretic Separation of the Enantiomers of Weak Acids with Hydroxypropyl. beta.-Cyclodextrin-Containing Background Electrolytes. *Analytical Chemistry* 66 (5):619-627.

[61] Rawjee, Yasir Y, Robert L Williams, Lisa A Buckingham, and Gyula Vigh. (1994). Effects of pH and hydroxypropyl β-cyclodextrin concentration on peak resolution in the capillary electrophoretic separation of the enantiomers of weak bases. *Journal of Chromatography A* 688 (1-2):273-282.

[62] Rawjee, Yasir Y, Robert L Williams, and Gyula Vigh. (1994). Capillary electrophoretic chiral separations using cyclodextrin additives: III. Peak resolution surfaces for ibuprofen and homatropine as a function of the pH and the concentration of β-cyclodextrin. *Journal of Chromatography A* 680 (2):599-607.

[63] Szücs, Roman, Isabelle Caron, Karen A Taylor, Steve P Gee, Paul D Ferguson, Martin A Kelly, Jon V Beaman, Andrew M Lipczynski, and Perry A Hailey. (2000). Generic approach to chiral separations: Chiral capillary electrophoresis with ternary cyclodextrin mixtures. *Journal of Microcolumn Separations* 12 (11):568-576.

[64] Zhou, Lili, Richard Thompson, Sherry Song, Dean Ellison, and Jean M Wyvratt. (2002). A strategic approach to the development of capillary electrophoresis chiral methods for pharmaceutical basic compounds using sulfated cyclodextrins. *Journal of pharmaceutical and biomedical analysis* 27 (3-4):541-553.

[65] Jimidar, M Ilias, Willy Van Ael, Patrick Van Nyen, Margot Peeters, Dirk Redlich, and Maurits De Smet. (2004). A screening strategy for the development of enantiomeric separation methods in capillary electrophoresis. *Electrophoresis* 25 (16):2772-2785.

[66] Sokoließ, Torsten, and Gerhard Köller. (2005). Approach to method development and validation in capillary electrophoresis for enantiomeric purity testing of active basic pharmaceutical ingredients. *Electrophoresis* 26 (12):2330-2341.

[67] Sänger–van de Griend, Cari E, Ylva Hedeland, and Curt Pettersson. (2013). Capillary Electrophoresis: an Attractive Technique for Chiral Separations. *Chromatography Today* 6 (2):32-37.

[68] Altria, KD. (1993). Essential peak area normalisation for quantitative impurity content determination by capillary electrophoresis. *Chromatographia* 35 (3-4):177-182.

[69] Owens, Paul K, Håkan Wikström, Sofia Någård, and Lars Karlsson. (2002). Development and validation of a capillary electrophoresis method for ximelagatran assay and related substance determination in drug substance and tablets. *Journal of pharmaceutical and biomedical analysis* 27 (3-4):587-598.

[70] de l'Escaille, François, and Jean-Bernard Falmagne. (2008). 13 ion analysis using capillary electrophoresis. *Separation Science and Technology* 9:317-355.

[71] Paulekuhn, G Steffen, Jennifer B Dressman, and Christoph Saal. (2007). Trends in active pharmaceutical ingredient salt selection based on analysis of the Orange Book database. *Journal of medicinal chemistry* 50 (26):6665-6672.

[72] Elbashir, Abdalla A, and Hassan Y Aboul-Enein. (2014). Recent applications and developments of capacitively coupled contactless conductivity detection (CE-C4D) in capillary electrophoresis. *Biomedical Chromatography* 28 (11):1502-1506.

[73] Štěpánová, Sille, and Václav Kašička. (2014). Determination of impurities and counterions of pharmaceuticals by capillary electromigration methods. *Journal of separation science* 37 (15):2039-2055.

[74] Timerbaev, Andrei R. (2002.) Recent advances and trends in capillary electrophoresis of inorganic ions. *Electrophoresis* 23 (22-23):3884-3906.

[75] Altria, KD, DM Goodall, and MM Rogan. (1994). Quantitative determination of drug counter-ion stoichiometry by capillary electrophoresis. *Chromatographia* 38 (9-10):637-642.

[76] Altria, KD, NG Clayton, RC Harden, JV Makwana, and MJ Portsmouth. (1995). Inter-company cross validation exercise on capillary electrophoresis. Quantitative determination of drug counter-ion level. *Chromatographia* 40 (1-2):47-50.

[77] Maruszak, Wioleta, and Marcin Cybulski. (2015). Simultaneous determination of the counter ion and possible impurity from the synthetic route in the pharmaceutical substance prasugrel hydrochloride. *Journal of pharmaceutical and biomedical analysis* 105:32-38.

[78] Stålberg, Olle, Karin Sander, and Cari Sänger-van De Griend. (2002). The determination of bromide in a local anaesthetic hydrochloride by capillary electrophoresis using direct UV detection. *Journal of Chromatography A* 977 (2):265-275.

[79] Zhang, Lei, Shen Luo, and Baolin Zhang. (2016). Glycan analysis of therapeutic glycoproteins. *MAbs.*

[80] Guttman, András. (2013). Capillary electrophoresis in the N-glycosylation analysis of biopharmaceuticals. *TrAC Trends in Analytical Chemistry* 48:132-143.

[81] Aich, Udayanath, Jude Lakbub, and Aston Liu. (2016). State-of-the-art technologies for rapid and high-throughput sample preparation and analysis of N-glycans from antibodies. *Electrophoresis* 37 (11):1468-1488.

[82] Volpi, Nicola. (2010). *Capillary electrophoresis of carbohydrates: from monosaccharides to complex polysaccharides*: Springer Science & Business Media.

[83] Rovio, Stella, Jari Yli-Kauhaluoma, and Heli Sirén. (2007). Determination of neutral carbohydrates by CZE with direct UV detection. *Electrophoresis* 28 (17):3129-3135.

[84] Analis. "*Application of Capillary Electrophoresis in Optimization and Control of Fermentation and Cell Culture.*" http://www.analis.be/site/objects/media/0/0/5/1/1/0051128_media/media1.pdf.

[85] Analis. "*Easy Method to Analyze Carbohydrates using Capillary Electrophoresis and direct UV detection.*" http://www.analis.be/site/objects/media/0/0/7/3/6/0073661_media/media1.pdf.

[86] Guttman, Andras, Marta Kerekgyarto, and Gabor Jarvas. (2015). Effect of Separation Temperature on Structure Specific Glycan Migration in Capillary Electrophoresis. *Analytical chemistry* 87 (23):11630-11634.

[87] Szekrényes, Ákos, SungAe Suhr Park, Marcia Santos, Clarence Lew, Aled Jones, Ted Haxo, Michael Kimzey, Shiva Pourkaveh, Zoltán Szabó, and Zoran Sosic. (2016). Multi-Site N-glycan mapping study 1: Capillary electrophoresis–laser induced fluorescence. *MAbs.*

[88] Reusch, Dietmar, Markus Haberger, Tobias Kailich, Anna-Katharina Heidenreich, Michael Kampe, Patrick Bulau, and Manfred Wuhrer. (2014). *"High-throughput glycosylation analysis of therapeutic immunoglobulin G by capillary gel electrophoresis using a DNA analyzer."* MAbs.

[89] Gahoual, Rabah, Alicia Burr, Jean-Marc Busnel, Lauriane Kuhn, Phillipe Hammann, Alain Beck, Yannis-Nicolas François, and Emmanuelle Leize-Wagner. (2013). *Rapid and multi-level characterization of trastuzumab using sheathless capillary electrophoresis-tandem mass spectrometry*. MAbs.

[90] Gahoual, Rabah, Alain Beck, Emmanuelle Leize-Wagner, and Yannis-Nicolas François. (2016). Cutting-edge capillary electrophoresis characterization of monoclonal antibodies and related products. *Journal of Chromatography B* 1032:61-78.

[91] Identification section B of the erythropoietin concentrated solution monograph. *PharmEuropa* 11 (1999):414-415.

[92] Bristow, A, and E Charton. (1999). Assessment of the suitability of a Capillary Zone Electrophoresis method for determining Isoform Distribution of Erythropoietin. *Pharmeuropa* 11 (2):290-300.

[93] Zhang, Le, Ken Lawson, Bernice Yeung, and Jette Wypych. (2014). Capillary zone electrophoresis method for a highly glycosylated and sialylated recombinant protein: development, characterization and

application for process development. *Analytical chemistry* 87 (1):470-476.

[94] Haselberg, Rob, Gerhardus J de Jong, and Govert W Somsen. (2013). CE-MS for the analysis of intact proteins 2010–2012. *Electrophoresis* 34 (1):99-112.

[95] Dotz, Viktoria, Rob Haselberg, Archana Shubhakar, Radoslaw P Kozak, David Falck, Yoann Rombouts, Dietmar Reusch, Govert W Somsen, Daryl L Fernandes, and Manfred Wuhrer. (2015). Mass spectrometry for glycosylation analysis of biopharmaceuticals. *TrAC Trends in Analytical Chemistry* 73:1-9.

[96] Haselberg, Rob, and Govert W Somsen. (2016). CE-MS for the analysis of intact proteins. *Capillary Electrophoresis-Mass Spectrometry (CE-MS): Principles and Applications*.

[97] Toppazzini, Mila, Anna Coslovi, and Sergio Paoletti. (2011). Capillary Electrophoresis Applied to Polysaccharide Characterization. In *Capillary Electrophoresis of Carbohydrates*, 127-149. Springer.

[98] Ramawat, Kishan Gopal, and Jean-Michel Mérillon. (2015). *Polysaccharides: bioactivity and biotechnology*: Springer Heidelberg.

[99] L Simeone, M Toppazzini, S Ricci, M Morelli, B Fabbri, G Rosy, S Giannini, C Campa. (2015). Approaches for Analysis of Unconjugated Components in GBS Glycoconjugate Vaccine: From Chromatography to Capillary Electrophoresis, presented at *CEPharm*, Brooklyn, NY.

[100] Görög, Sándor. (2017). Critical review of reports on impurity and degradation product profiling in the last decade. *TrAC Trends in Analytical Chemistry*.

[101] Szajek, Anita Y, Edward Chess, Kristian Johansen, Gyöngyi Gratzl, Elaine Gray, David Keire, Robert J Linhardt, Jian Liu, Tina Morris, and Barbara Mulloy. (2016). The US regulatory and pharmacopeia response to the global heparin contamination crisis. *Nature biotechnology* 34 (6):625.

[102] Somsen, Govert W, Yvonne H Tak, Javier Sastre Toraño, Peter MJM Jongen, and Gerhardus J de Jong. (2009). Determination of

oversulfated chondroitin sulfate and dermatan sulfate impurities in heparin by capillary electrophoresis. *Journal of Chromatography A* 1216 (18):4107-4112.

[103] Hörmann, Aldo, Maria Schwarz, and Fabian Stapf. (2011). Analyzing for Small Molecules in Large-Molecule Pharmaceuticals/The potential of capillary electrophoresis. *Pharmazeutische Industrie* 73 (12):2222.

[104] van de Goor, Tom AAM, Peter SL Janssen, Jan W Van Nispen, Mario JM van Zeeland, and Frans M Everaerts. (1991). Capillary electrophoresis of peptides: Analysis of adrenocorticotropic hormone-related fragments. *Journal of Chromatography A* 545 (2):379-389.

[105] Brunnkvist, Helena, Bo Karlberg, Åsa Astervik, and Ingrid Granelli. (2004). Experimental design-based development of a rapid capillary electrophoresis method for determining impurities in the tetrapeptide H-Tyr-(D) Arg-Phe-Phe-NH2. *Journal of Chromatography B* 807 (2):293-300.

[106] Gennaro, Lynn A, and Oscar Salas-Solano. (2009). Characterization of deamidated peptide variants by micro-preparative capillary electrophoresis and mass spectrometry. *Journal of Chromatography A* 1216 (20):4499-4503.

[107] Rush, Robert S, Patricia L Derby, Thomas W Strickland, and Michael F Rohde. (1993). Peptide mapping and evaluation of glycopeptide microheterogeneity derived from endoproteinase digestion of erythropoietin by affinity high-performance capillary electrophoresis. *Analytical chemistry* 65 (14):1834-1842.

[108] Yuan, Huimin, George M Janini, Haleem J Issaq, Richard A Thompson, and Dean K Ellison. (2000). *Separation of closely related heptadecapeptides by micellar electrokinetic chromatography.*

[109] Larsson, Marita, and ES Mareike Lutz. (2000). Transient isotachophoresis for sensitivity enhancement in capillary electrophoresis-mass spectrometry for peptide analysis. *Electrophoresis: An International Journal* 21 (14):2859-2865.

[110] Whitmore, Colin D, and Lynn A Gennaro. (2012). Capillary electrophoresis-mass spectrometry methods for tryptic peptide

mapping of therapeutic antibodies. *Electrophoresis* 33 (11):1550-1556.

[111] Wenz, Christian, Coral Barbas, Ángeles López-Gonzálvez, Antonia Garcia, Fernando Benavente, Victoria Sanz-Nebot, Tim Blanc, Gordon Freckleton, Philip Britz-McKibbin, and Meera Shanmuganathan. (2015). Interlaboratory study to evaluate the robustness of capillary electrophoresis-mass spectrometry for peptide mapping. *Journal of separation science* 38 (18):3262-3270.

[112] Lauer, HH, and D McManigill. (1986). Zone electrophoresis in open-tubular capillaries—recent advances. *TrAC Trends in Analytical Chemistry* 5 (1):11-15.

[113] Lauer, Henk H, and Douglass McManigill. (1986). Capillary zone electrophoresis of proteins in untreated fused silica tubing. *Analytical Chemistry* 58 (1):166-170.

[114] Frenz, John, Shiaw-Lin Wu, and William S Hancock. (1989). Characterization of human growth hormone by capillary electrophoresis. *Journal of Chromatography A* 480:379-391.

[115] Mandrup, Gitte. (1992). Rugged method for the determination of deamidation products in insulin solutions by free zone capillary electrophoresis using an untreated fused-silica capillary. *Journal of Chromatography A* 604 (2):267-281.

[116] Watson, E, and F Yao. (1993). Capillary electrophoretic separation of human recombinant erythropoietin (r-HuEPO) glycoforms. *Analytical biochemistry* 210 (2):389-393.

[117] Tran, An D, Sungae Park, Peter J Lisi, Oanh T Huynh, Ralph R Ryall, and Philip A Lane. (1991). Separation of carbohydrate-mediated microheterogeneity of recombinant human erythropoietin by free solution capillary electrophoresis: Effects of pH, buffer type and organic additives. *Journal of Chromatography A* 542:459-471.

[118] Ma, S, and W Nashabeh. (2001). Analysis of protein therapeutics by capillary electrophoresis. In *CE in Biotechnology: Practical Applications for Protein and Peptide Analyses*, 75-89. Springer.

[119] "*PubChem Open Chemistry Database.*" https://pubchem.ncbi.nlm.nih.gov/ compound/ 6-aminohexanoic_acid#section=Chemical-and-Physical-Properties.

[120] Green, Jonathan S, and James W Jorgenson. (1989). Minimizing adsorption of proteins on fused silica in capillary zone electrophoresis by the addition of alkali metal salts to the buffers. *Journal of Chromatography A* 478:63-70.

[121] Wätzig, Hermann, Matthias Degenhardt, and Annette Kunkel. (1998). Strategies for capillary electrophoresis: method development and validation for pharmaceutical and biological applications. *Electrophoresis* 19 (16-17):2695-2752.

[122] He, Yan, Nathan A Lacher, Weiying Hou, Qian Wang, Colleen Isele, Jason Starkey, and Margaret Ruesch. (2010). Analysis of identity, charge variants, and disulfide isomers of monoclonal antibodies with capillary zone electrophoresis in an uncoated capillary column. *Analytical chemistry* 82 (8):3222-3230.

[123] Fung, Eliza N, and Edward S Yeung. (1995). High-speed DNA sequencing by using mixed poly (ethylene oxide) solutions in uncoated capillary columns. *Analytical Chemistry* 67 (13):1913-1919.

[124] Nevins, Stacey A, Barbara A Siles, and Zeena E Nackerdien. (2000). Analysis of gamma radiation-induced damage to plasmid DNA using dynamic size-sieving capillary electrophoresis. *Journal of Chromatography B: Biomedical Sciences and Applications* 741 (2):243-255.

[125] Deng, AY, Byrne, MP, Spitznagel, TM, Wu, Z. (2006). Presented at *CEPharm*, Jersey City, NJ.

[126] Shi, Ying, Zhen Li, Yuanbiao Qiao, and Jun Lin. (2012). Development and validation of a rapid capillary zone electrophoresis method for determining charge variants of mAb. *Journal of Chromatography B* 906:63-68.

[127] Li, Y., MP Reddy, CK Ratnayake, EV Koh. (2009). *Methods and compositions for capillary electrophoresis,* Patent application number 20090314638.

[128] Andrews, AT, and CF Simpson. (1987). *Electrophoresis: Theory, techniques and biochemical and clinical applications,* 2nd ed.: Oxford Science Publications, Monographs on Physical Chemistry, Clarendon Press, Oxford, 1986 (ISBN 0-19-854633-5 hard back or-854632-7 soft back). xv+ 452 pp. Price£ 30.00 (hard back),£ 15.00 (soft back). Elsevier.

[129] Michels, David A, Monica Parker, and Oscar Salas-Solano. (2012). Quantitative impurity analysis of monoclonal antibody size heterogeneity by CE-LIF: Example of development and validation through a quality-by-design framework. *Electrophoresis* 33 (5):815-826.

[130] F Winkhaus, M Haindl. (2016). Microchip-CE: smaller is faster. *Analytical Scientist* (2):18 – 20.

[131] Smith, Michael T, Shu Zhang, Troy Adams, Byron DiPaolo, and Jennifer Dally. (2017). Establishment and validation of a microfluidic capillary gel electrophoresis platform method for purity analysis of therapeutic monoclonal antibodies. *Electrophoresis* 38 (9-10):1353-1365.

[132] Chen, Grace, Sha Ha, and Richard R Rustandi. (2013). Characterization of glycoprotein biopharmaceutical products by Caliper LC90 CE-SDS Gel Technology. In *Glycosylation Engineering of Biopharmaceuticals*, 199-209. Springer.

[133] Le, M Eleanor, Alona Vizel, and Katariina M Hutterer. (2013). Automated sample preparation for CE-SDS. *Electrophoresis* 34 (9-10):1369-1374.

[134] Paul, Suparna, Judy Connor, Tom Nesspor, Peter Haytko, Ken Boakye, Mark L Chiu, and Haiyan Jiang. (2016). An efficient process of generating bispecific antibodies via controlled Fab-arm exchange using culture supernatants. *Protein expression and purification* 121:133-140.

[135] Chen, Tao, Yan Chen, Cinzia Stella, Colin D Medley, Jason A Gruenhagen, and Kelly Zhang. (2016). Antibody-drug conjugate characterization by chromatographic and electrophoretic techniques. *Journal of Chromatography B* 1032:39-50.

[136] van Tricht, Ewoud, Lars Geurink, Bojana Pajic, Johan Nijenhuis, Harold Backus, Marta Germano, Govert W Somsen, and Cari E Sänger-van de Griend. (2015). New capillary gel electrophoresis method for fast and accurate identification and quantification of multiple viral proteins in influenza vaccines. *Talanta* 144:1030-1035.

[137] Sánchez-Hernández, Laura, Cristina Montealegre, Steffen Kiessig, Bernd Moritz, and Christian Neusüß. (2017). In-capillary approach to eliminate SDS interferences in antibody analysis by capillary electrophoresis coupled to mass spectrometry. *Electrophoresis* 38 (7):1044-1052.

[138] Hjerten, Stellan. (1985). Adaptation of the equipment for high-performance electrophoresis to isoelectric focusing. *Journal of Chromatography A* 346:265-270.

[139] Hjerten, Stellan, Karin Elenbring, Ferenc Kilár, Jia-Li Liao, Albert JC Chen, Christopher J Siebert, and Ming-De Zhu. (1987). Carrier-free zone electrophoresis, displacement electrophoresis and isoelectric focusing in a high-performance electrophoresis apparatus. *Journal of Chromatography A* 403:47-61.

[140] Mosher, Richard A, Wolfgang Thormann, Aly Graham, and Milan Bier. (1985). The formation of stable pH gradients with weak monovalent buffers for isoelectric focusing in free solution. *Electrophoresis* 6 (11):545-551.

[141] Thormann, Wolfgang, Richard A Mosher, and Milan Bier. (1986). Experimental and theroretical dynamics of isoelectric focusing: Elucidation of a general separation mechanism. *Journal of Chromatography A* 351:17-29.

[142] Mack, Scott, Ingrid Cruzado-Park, Jeff Chapman, Chitra Ratnayake, and Gyula Vigh. (2009). A systematic study in CIEF: defining and optimizing experimental parameters critical to method reproducibility and robustness. *Electrophoresis* 30 (23):4049-4058.

[143] Wu, Jiaqi, and Janusz Pawliszyn. (1992). Universal detection for capillary isoelectric focusing without mobilization using concentration gradient imaging system. *Analytical Chemistry* 64 (2):224-227.

[144] Wu, Jiaqi, Shun-Cheng Li, and Arthur Watson. (1998). Optimizing separation conditions for proteins and peptides using imaged capillary isoelectric focusing. *Journal of Chromatography A* 817 (1-2):163-171.

[145] Loughney, John W, Sha Ha, and Richard R Rustandi. (2017). Quantitation of CRM197 using imaged capillary isoelectric focusing with fluorescence detection and capillary Western. *Analytical biochemistry* 534:19-23.

[146] Zarabadi, Atefeh S, Tiemin Huang, and John G Mielke. (2017). Capillary isoelectric focusing with whole column imaging detection (iCIEF): A new approach to the characterization and quantification of salivary α-amylase. *Journal of Chromatography B* 1053:65-71.

[147] Wenz, C. (2017). "*Comparison of Different Brands of Carrier Ampholytes for Monoclonal Antibody Charge Heterogeneity Analysis by Capillary Isoelectric Focusing*." https://www.semanticscholar.org/paper/Comparison-of-Different-Brands-of-Carrier-for-by-Wenz/0abd2c5b94c938e857031c3f5c7397ead1f51edb ?tab=abstract.

[148] Hühner, Jens, Michael Lämmerhofer, and Christian Neusüß. (2015). Capillary isoelectric focusing-mass spectrometry: Coupling strategies and applications. *Electrophoresis* 36 (21-22):2670-2686.

[149] Dai, Jun, Jared Lamp, Qiangwei Xia, and Yingru Zhang. (2018). Capillary Isoelectric Focusing-Mass Spectrometry Method for the Separation and Online Characterization of Intact Monoclonal Antibody Charge Variants. *Analytical chemistry* 90 (3):2246-2254.

[150] Lin, Shiming, and Su-Ming Hsu. (1997). Determination of the association constant for slow phosphoserine-anti-phosphoserine interaction kinetics by capillary zone electrophoresis. *Electrophoresis* 18 (11):2042-2046.

[151] Anderson, Carrie L, Yang Wang, and Richard R Rustandi. (2012). Applications of imaged capillary isoelectric focussing technique in development of biopharmaceutical glycoprotein-based products. *Electrophoresis* 33 (11):1538-1544.

[152] Heegaard, Niels HH, Staffan Nilsson, and Norberto A Guzman. (1998). Affinity capillary electrophoresis: important application areas

and some recent developments. *Journal of Chromatography B: Biomedical Sciences and Applications* 715 (1):29-54.

[153] Guzman, Norberto A, and R John Stubbs. (2001). The use of selective adsorbents in capillary electrophoresis-mass spectrometry for analyte preconcentration and microreactions: A powerful three-dimensional tool for multiple chemical and biological applications. *Electrophoresis* 22 (17):3602-3628.

[154] Rustandi, Richard R, John W Loughney, Melissa Hamm, Christopher Hamm, Catherine Lancaster, Anna Mach, and Sha Ha. (2012). Qualitative and quantitative evaluation of Simon™, a new CE-based automated Western blot system as applied to vaccine development. *Electrophoresis* 33 (17):2790-2797.

[155] d'Orlyé, Fanny, Anne Varenne, and Pierre Gareil. (2008). Determination of nanoparticle diffusion coefficients by Taylor dispersion analysis using a capillary electrophoresis instrument. *Journal of Chromatography A* 1204 (2):226-232.

[156] Østergaard, Jesper, and Henrik Jensen. (2009). Simultaneous evaluation of ligand binding properties and protein size by electrophoresis and Taylor dispersion in capillaries. *Analytical chemistry* 81 (20):8644-8648.

[157] Hawe, Andrea, Wendy L Hulse, Wim Jiskoot, and Robert T Forbes. (2011). Taylor dispersion analysis compared to dynamic light scattering for the size analysis of therapeutic peptides and proteins and their aggregates. *Pharmaceutical research* 28 (9):2302-2310.

[158] Jensen, Sabrine S, Henrik Jensen, Claus Cornett, Eva H Møller, and Jesper Østergaard. (2014). Insulin diffusion and self-association characterized by real-time UV imaging and Taylor dispersion analysis. *Journal of pharmaceutical and biomedical analysis* 92:203-210.

[159] Poulsen, Nicklas N, Nina Z Andersen, Jesper Østergaard, Guisheng Zhuang, Nickolaj J Petersen, and Henrik Jensen. (2015). Flow induced dispersion analysis rapidly quantifies proteins in human plasma samples. *Analyst* 140 (13):4365-4369.

[160] Poulsen, Nicklas N, Morten E Pedersen, Jesper Østergaard, Nickolaj J Petersen, Christoffer T Nielsen, Niels HH Heegaard, and Henrik Jensen. (2016). Flow-induced dispersion analysis for probing anti-dsDNA antibody binding heterogeneity in systemic lupus erythematosus patients: toward a new approach for diagnosis and patient stratification. *Analytical chemistry* 88 (18):9056-9061.

[161] Barmé, Iris, Gerard JM Bruin, Aran Paulus, and Markus Ehrat. (1998). Preconcentration and separation of antisense oligonucleotides by on-column isotachophoresis and capillary electrophoresis in polymer-filled capillaries. *Electrophoresis* 19 (8-9):1445-1451.

[162] Orchard, MJ, IP Collin, and M Bushell. (1999). The use of capillary electrophoresis with entangled polymer matrix to analyse plasmid DNA and a cationic lipid/cholesterol liposome: DNA complex. *Chromatographia* 50 (3-4):181-187.

[163] Holovics, Heidi J, Yan He, Nathan A Lacher, and Margaret N Ruesch. (2010). Capillary gel electrophoresis with laser-induced fluorescence of plasmid DNA in untreated capillary. *Electrophoresis* 31 (14):2436-2441.

[164] Shambaugh, CL, J-L Bodmer, D Hsu, and CS Ranucci. (2004). Separation and quantification of viral double-stranded RNA fragments by capillary electrophoresis in hydroxyethylcellulose polymer solutions. *Journal of Chromatography A* 1051 (1-2):161-170.

[165] Palm, Anders K, and György Marko-Varga. (2004). On-column electroextraction and separation of antisense oligonucleotides in human plasma by capillary gel electrophoresis. *Journal of pharmaceutical and biomedical analysis* 35 (3):415-423.

[166] Székely, László, Steffen Kiessig, Maria A Schwarz, and Franka Kálmán. (2009). Capillary gel electrophoresis of therapeutic oligonucleotides–Analysis of single-and double-stranded forms. *Electrophoresis* 30 (9):1579-1586.

[167] Fourtounis, Jimmy, Jean-Pierre Falgueyret, and Camil Elie Sayegh. (2011). Assessing protein–RNA interactions using microfluidic capillary mobility shift assays. *Analytical biochemistry* 411 (1):161-163.

[168] Desai, Meera J, and Daniel W Armstrong. (2003). Separation, identification, and characterization of microorganisms by capillary electrophoresis. *Microbiology and molecular biology reviews* 67 (1):38-51.

[169] Rodriguez, Michael A, and Daniel W Armstrong. (2004). Separation and analysis of colloidal/nano-particles including microorganisms by capillary electrophoresis: a fundamental review. *Journal of Chromatography B* 800 (1-2):7-25.

[170] Kremser, Leopold, Dieter Blaas, and Ernst Kenndler. (2004). Capillary electrophoresis of biological particles: viruses, bacteria, and eukaryotic cells. *Electrophoresis* 25 (14):2282-2291.

[171] Schnabel, Ursula, Franz Groiss, Dieter Blaas, and Ernst Kenndler. (1996). Determination of the p I of Human Rhinovirus Serotype 2 by Capillary Isoelectric Focusing. *Analytical chemistry* 68 (23):4300-4303.

[172] Okun, Vadim M, Bernhard Ronacher, Dieter Blaas, and Ernst Kenndler. (2000). Affinity capillary electrophoresis for the assessment of complex formation between viruses and monoclonal antibodies. *Analytical chemistry* 72 (19):4634-4639.

[173] Okun, Vadim M, Bernhard Ronacher, Dieter Blaas, and Ernst Kenndler. (1999). Analysis of common cold virus (human rhinovirus serotype 2) by capillary zone electrophoresis: the problem of peak identification. *Analytical chemistry* 71 (10):2028-2032.

[174] Oita, Iuliana, Hadewych Halewyck, Sigrid Pieters, Bieke Dejaegher, Bert Thys, Bart Rombaut, and Yvan Vander Heyden. (2009). Improving the capillary electrophoretic analysis of poliovirus using a Plackett–Burman design. *Journal of pharmaceutical and biomedical analysis* 50 (4):655-663.

[175] Halewyck, Hadewych, Iuliana Oita, Bert Thys, Bieke Dejaegher, Yvan Vander Heyden, and Bart Rombaut. (2010). Identification of poliovirions and subviral particles by capillary electrophoresis. *Electrophoresis* 31 (19):3281-3287.

[176] Oita, Iuliana, Hadewych Halewyck, Sigrid Pieters, Bert Thys, Yvan Vander Heyden, and Bart Rombaut. (2012). Poliovirus separation

from cell extracts using capillary electrophoresis: Potential use in vaccine production and control? *Journal of virological methods* 185 (1):7-17.

[177] Oita, Iuliana, Hadewych Halewyck, Sigrid Pieters, Bieke Dejaegher, Bert Thys, Bart Rombaut, and Yvan Vander Heyden. (2011). Rational use of stacking principles for signal enhancement in capillary electrophoretic separations of poliovirus samples. *Journal of pharmaceutical and biomedical analysis* 55 (1):135-145.

[178] Oita, Iuliana, Hadewych Halewyck, Bert Thys, Bart Rombaut, and Yvan Vander Heyden. (2012). Impact of the sample matrix composition on the signal enhancement in the capillary electrophoretic separation of poliovirus samples. *Analytica chimica acta* 747:42-50.

[179] Oita, Iuliana, Hadewych Halewyck, Bert Thys, Bart Rombaut, and Yvan Vander Heyden. (2012). The effect of anionic surfactant on poliovirus particles during capillary electrophoresis. *Journal of pharmaceutical and biomedical analysis* 71:79-88.

[180] Halewyck, Hadewych, Lise Schotte, Iuliana Oita, Bert Thys, Ann Van Eeckhaut, Yvan Vander Heyden, and Bart Rombaut. (2014). Affinity capillary electrophoresis to evaluate the complex formation between poliovirus and nanobodies. *Journal of separation science* 37 (24):3729-3737.

[181] Bettonville, Virginie, Jérôme TJ Nicol, Nicolas Thelen, Marc Thiry, Marianne Fillet, Nathalie Jacobs, and Anne-Catherine Servais. (2016). Study of intact virus-like particles of human papillomavirus by capillary electrophoresis. *Electrophoresis* 37 (4):579-586.

[182] Bettonville, Virginie, Jerome TJ Nicol, Tania Furst, Nicolas Thelen, Géraldine Piel, Marc Thiry, Marianne Fillet, Nathalie Jacobs, and Anne-Catherine Servais. (2017). Quantitation and biospecific identification of virus-like particles of human papillomavirus by capillary electrophoresis. *Talanta* 175:325-330.

[183] Mann, Bruce, Joseph A Traina, Cynthia Soderblom, Peter K Murakami, Elisabeth Lehmberg, Dennis Lee, John Irving, Eirik Nestaas, and Erno Pungor Jr. (2000). Capillary zone electrophoresis

of a recombinant adenovirus. *Journal of Chromatography A* 895 (1-2):329-337.

[184] van Tricht, Ewoud, Lars Geurink, Harold Backus, Marta Germano, Govert W Somsen, and Cari E Sänger–van de Griend. (2017). One single, fast and robust capillary electrophoresis method for the direct quantification of intact adenovirus particles in upstream and downstream processing samples. *Talanta* 166:8-14.

[185] Kundu, S, C Fenters, M Lopez, A Varma, J Brackett, S Kuemmerle, and JC Hunt. (1997). Capillary electrophoresis for purity estimation and in-process testing of recombinant GB virus-C proteins. *Journal of capillary electrophoresis* 4 (1):7-13.

[186] Schwarzer, Jana, Erdmann Rapp, and Udo Reichl. (2008). N-glycan analysis by CGE–LIF: Profiling influenza A virus hemagglutinin N-glycosylation during vaccine production. *Electrophoresis* 29 (20):4203-4214.

[187] Mellado, Maria Candida M, Cristina Peixoto, Pedro E Cruz, Manuel JT Carrondo, and Paula M Alves. (2008). Purification of recombinant rotavirus VP7 glycoprotein for the study of in vitro rotavirus-like particles assembly. *Journal of Chromatography B* 874 (1-2):89-94.

[188] Wright, Peter F. (2007). Orthomyxoviruses. *Fields virology*:1691-1740.

[189] Gollapudi, Deepika, Diane L Wycuff, Richard M Schwartz, Jonathan W Cooper, and KC Cheng. (2017). Development of high-throughput and high sensitivity capillary gel electrophoresis platform method for Western, Eastern, and Venezuelan equine encephalitis (WEVEE) virus like particles (VLPs) purity determination and characterization. *Electrophoresis* 38 (20):2610-2621.

[190] Gullmar, B, S Hjerten, and T Wadström. (1988). Cell surface charge of lactobacilli and enterococci isolated from pig small intestine as studied by free zone electrophoresis: a methodological study. *Microbios* 55 (224-225):183-192.

[191] Lantz, Andrew W, Ye Bao, and Daniel W Armstrong. (2007). Single-cell detection: test of microbial contamination using capillary electrophoresis. *Analytical chemistry* 79 (4):1720-1724.

[192] Bao, Jiayin, Svetlana M Krylova, Leonid T Cherney, Robert L Hale, Svetlana L Belyanskaya, Cynthia H Chiu, Christopher C Arico-Muendel, and Sergey N Krylov. (2015). Prediction of protein–DNA complex mobility in gel-free capillary electrophoresis. *Analytical chemistry* 87 (4):2474-2479.

[193] Reid, George L, James Morgado, Kimber Barnett, Brent Harrington, Jian Wang, J Harwood, and D Fortin. (2013). Analytical Quality by Design (AQbD) in pharmaceutical development. *American Pharmaceutical Review* 144191.

[194] Peraman, Ramalingam, Kalva Bhadraya, and Yiragamreddy Padmanabha Reddy. (2015). Analytical quality by design: a tool for regulatory flexibility and robust analytics. *International journal of Analytical chemistry* 2015.

[195] Martin, Gregory P, Kimber L Barnett, Christopher Burgess, Paul D Curry, Joachim Ermer, GS Gratzl, JP Hammond, J Herrmann, E Kovacs, and DJ LeBlond. (2017). *"Proposed New USP General Chapter: The Analytical Procedure Lifecycle (1220)."* US Pharmacopeial Convention. Web.

[196] Niedermeier, Stephan, and Gerhard KE Scriba. (2017). A quality by design-based approach to a capillary electrokinetic assay for the determination of dextromepromazine and levomepromazine sulfoxide as impurities of levomepromazine. *Journal of pharmaceutical and biomedical analysis* 146:402-409.

[197] Krait, Sulaiman, Matthias Heuermann, and Gerhard KE Scriba. (2017). Development of a capillary electrophoresis method for the determination of the chiral purity of dextromethorphan by a dual selector system using quality by design methodology. *Journal of separation science*.

[198] Arabi, Maryam, Mehrorang Ghaedi, and Abbas Ostovan. (2017). Development of a lower toxic approach based on green synthesis of water-compatible molecularly imprinted nanoparticles for the extraction of hydrochlorothiazide from human urine. *ACS Sustainable Chemistry & Engineering* 5 (5):3775-3785.

[199] Orlandini, Serena, Benedetta Pasquini, Claudia Caprini, Massimo Del Bubba, Sergio Pinzauti, and Sandra Furlanetto. (2015). Analytical Quality by Design in pharmaceutical quality assurance: Development of a capillary electrophoresis method for the analysis of zolmitriptan and its impurities. *Electrophoresis* 36 (21-22):2642-2649.

[200] Orlandini, S, B Pasquini, M Stocchero, S Pinzauti, and S Furlanetto. (2014). An integrated quality by design and mixture-process variable approach in the development of a capillary electrophoresis method for the analysis of almotriptan and its impurities. *Journal of Chromatography A* 1339:200-209.

ABOUT THE EDITORS

Dr. Oscar Núñez studied chemistry at the University of Barcelona (UB) where he also received his PhD in 2004. He worked as visiting researcher for half a year in the development of on-line preconcentration methods using micellar electrokinetic chromatography (MEKC) at the University of Hyogo (Japan) in collaboration with Professor Shigeru Terabe (father of MEKC technique). Starting in October 2005, he joined Kyoto Institute of Technology (Japan) as a two-year post-doc researcher working with Professor Nobuo Tanaka developing monolithic silica capillary columns under a fellowship from the Japan Society for the Promotion of Science. In November 2007 he joined UB again, working in the Chromatography, Capillary Electrophoresis and Mass Spectrometry research group. He has more than ninety scientific papers and book chapters to his name, and he is the editor of two books on LC-MS/MS and sample preparation techniques. He has extensive experience in the development of capillary electrophoresis, liquid chromatography, mass spectrometry and high resolution mass spectrometry methods, as well as sample treatment procedures for environmental and food analysis. Since 2010, his major research area include the characterization, classification and authentication of natural food products in the prevention of frauds by means of separation techniques, mass spectrometry and chemometrics.

Dr. Xavier Subirats received his degree in Chemistry and PhD (2007) at the University of Barcelona. His doctoral thesis dealt with modeling the retention of ionizable compounds in liquid chromatography by studying the pK_a and pH variations in hydroorganic mobile phases, and the investigation of the role of organic solvents in non-aqueous capillary electrophoresis (NACE). These latter studies were mainly performed in 2004 at the group of Prof. Ernst Kenndler at the University of Vienna (Austria). In 2009 he joined the research group of Dr. Dieter Blaas in the Department of Medical Biochemistry of the Max F. Perutz Laboratories (Vienna, Austria), using liposomes as model systems for lipid membranes and electrophoretic techniques (in capillary format and chip) for the study of the infection process of the human rhinovirus. At the end of 2011 he returned to Barcelona and worked in collaboration with the pharmaceutical company Almirall in the determination of lipophilicity by chromatographic techniques. In 2013 he worked in the division of active principal ingredients of Esteve group in order to develop analytical methods for the analysis of raw materials, intermediates, process controls and final products. He rejoined the University of Barcelona in 2014 with full teaching and research responsibilities, and since then his research lines are focused on the surrogation of biological systems using chromatographic and electrophoretic systems, in the determination of physicochemical properties of substances of pharmaceutical interest and in the modeling of chromatographic retention in HILIC.

INDEX

B

C

D

E

F

L

M

N

O

P

Q

R

S

T

U

V

W

X

Y

Z

Capillary Electrophoresis: Fundamentals, Techniques and Applications

Editor: Zhongqi He

Series: Chemistry Research and Applications

Book Description: Capillary electrophoresis (CE) is an analytical technique that separates ions based on their electrophoretic mobility with the use of an applied voltage.

Hardcover ISBN: 978-1-62081-785-8
Retail Price: $182

Capillary Electrophoresis (CE): Principles, Challenges and Applications

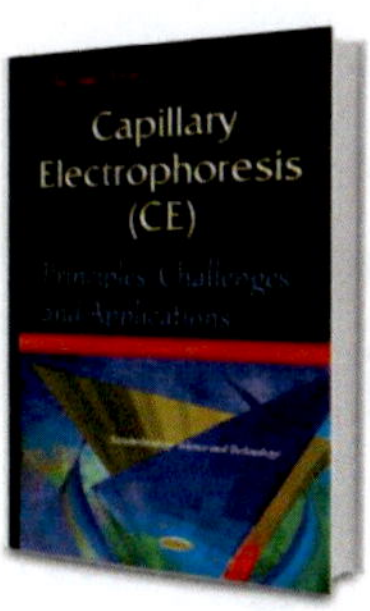

Editor: Christian Reed

Series: Nanotechnology Science and Technology

Book Description: This book examines challenges and applications, as well as principles of capillary electrophoresis.

Hardcover ISBN: 978-1-63483-122-2
Retail Price: $255

Handbook of Membrane Research: Properties, Performance and Applications

Editor: Stephan V. Gorley

Series: Chemical Engineering Methods and Technology

Book Description: Membranes can be defined as a selective barrier between two phases and classified into biological or synthetic membranes. This new and important book gathers the latest research from around the world in discussion of this dynamic field.

Hardcover ISBN: 978-1-60741-638-8
Retail Price: $350

Electrophoretic Deposition (EPD): Advances in Applications and Research

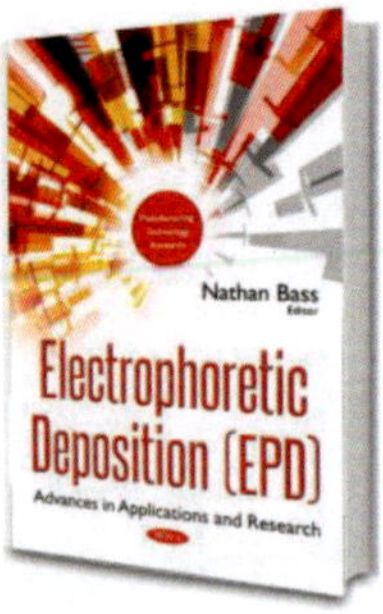

Editor: Nathan Bass

Series: Manufacturing Technology Research

Book Description: Electrophoretic deposition (EPD) is attracting many researchers' attention nowadays because of its numerous advantages, such as simple deposition apparatus, fast deposition rate, and the ease of deposition with a controlled thickness, compared to the other processing techniques.

Softcover ISBN: 978-1-53612-302-9
Retail Price: $82